VOL.1

Uncovering Student Ideas in Life Science

25 New Formative Assessment Probes

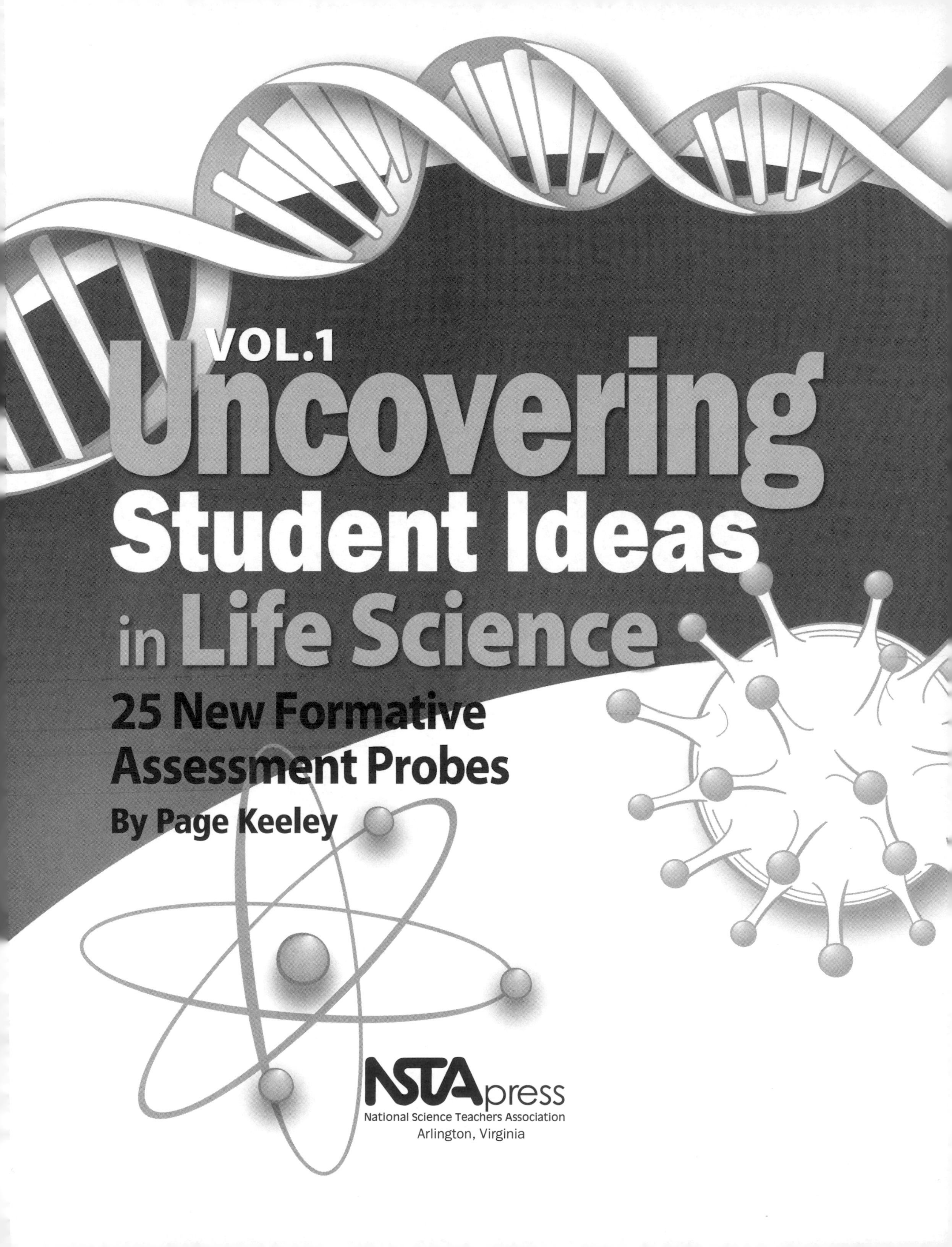

VOL. 1 Uncovering Student Ideas in Life Science

25 New Formative Assessment Probes

By Page Keeley

NSTApress
National Science Teachers Association
Arlington, Virginia

Claire Reinburg, Director
Jennifer Horak, Managing Editor
Andrew Cooke, Senior Editor
Judy Cusick, Senior Editor
Wendy Rubin, Associate Editor
Amy America, Book Acquisitions Coordinator

Art and Design
Will Thomas Jr., Director
Cover, Inside Design, and Illustrations by Linda Olliver

Printing and Production
Catherine Lorrain, Director

National Science Teachers Association
Francis Q. Eberle, PhD, Executive Director
David Beacom, Publisher
1840 Wilson Blvd., Arlington, VA 22201
www.nsta.org/store
For customer service inquiries, please call 800-277-5300.

14 13 12 11 5 4 3 2

Library of Congress Cataloging-in-Publication Data
Keeley, Page.
Uncovering student ideas in life science / by Page Keeley.
p. cm.
Includes bibliographical references and index.
ISBN 978-1-936137-17-6 (v. 1)
1. Life sciences—Study and teaching. 2. Educational evaluation. I. Title.
QH315.K37 2011
570.71--dc22
2010052477

eISBN 978-1-936137-51-0

Contents

Section 2

Ecosystems and Adaptation

Reproduction, Life Cycles, and Heredity

Human Biology

Dedication

This book is dedicated to Ray Barber, biology teacher extraordinaire at Pleasant Valley High School in Chico, California. He is a Toyota TAPESTRY awardee, friend, and colleague whom I have had the pleasure to get to know through NSTA and my work on formative assessment.

Ray, your innovation, creativity, and connection to your students truly inspire me! You exemplify what it means to be part of the science education community, a dedicated NSTA member, and a lifelong learner. Thanks for all you do to support students. And a huge thanks for giving NSTA Press authors like me a reason to continue to churn out these books for great teachers like you!

Foreword

Capturing the conceptual knowledge of students on a particular science topic is one of the most important things that science teachers do. With this knowledge, teachers comprehend the depth of understanding among students or their prevailing misconceptions. More important, by having this knowledge, teachers can determine what instructional strategies to use to ensure that students develop a clear understanding of a science topic. These strategies can range from asking students (verbally or in writing) to explain their thinking to having them draw conclusions about various phenomena. Before such instructional strategies can be put into place, however, teachers have to assess their students' knowledge.

Teachers have a multitude of ways to capture student knowledge. Tests, quizzes, questions, and looking at the written work of students are standard methods. A method used less frequently, but one that is equally important, is the formative assessment probe, which is the approach used in this book (and other books by Page Keeley and her coauthors; see p. xviii for a complete list of these books). Probes allow a science teacher to determine the depth of student understanding in specific conceptual areas. Making this determination is essential because even though a student may be able to select the correct answer to a question from several possible answers in a probe, he or she may not have a deep understanding of the topic, as would be demonstrated by the student's explanation for choosing that answer, a step that the probes require.

The probes in this book are simple yet useful indicators of student understanding. Each probe presents a scenario that involves selecting an answer and providing an explanation. The elegant format of the Teacher Notes that accompany each probe includes a discussion of research about student conceptual knowledge in the area targeted by the probe and the related National Science Education Standards (NRC 1996) and Benchmarks for Science Literacy (AAAS 1993, 2009). This connection of research to national science standards tells teachers what students are likely to misunderstand and what level of understanding students can be expected to have within a particular grade span. Few science education assessment resources offer this magic combination of being research-based, standards-linked, and aligned with topics regularly taught in the science classroom.

For me, another important use of these probes can be the connection of research and practice. Although the probes were designed to capture the preconceptions that students have, they can also be used as tools to monitor how students are thinking about a concept during and after classroom instruction. Teachers who monitor the thinking of their students, and draw conclusions about student thinking, are engaged in a research process that is connected to their classroom practices.

This connection of research to practice can even be used by *groups of teachers* to study student thinking before, during, and after instruction. With the probes, groups of science teachers within a school or district can collect data about students' thinking, come together, and compare the data across instructional topics in science and different curricular formats. When teachers work together on collecting and analyzing student data, they form communities that focus on learning about student thinking and understanding and they definitely enhance their practices.

Foreword

All teachers, preservice through experienced, will find these probes valuable. *Preservice teachers* can use the probes to understand students' thinking about natural phenomena. By considering how students think about scientific phenomena, preservice teachers can begin to understand how important it is to tailor their instruction to the preconceptions their students bring to their learning. *Beginning science teachers* who use these probes will build their capacity to examine student thinking and formalize their ability to capture student ideas and modify their instruction accordingly. *Experienced teachers* who use these probes will continue to develop their expertise in using student understandings to create interactive learning environments that are rich in dialogue and student explanations. In terms of these probes, we can safely say that one size does fit all teachers!

This book fills an important void in the *Uncovering Student Ideas in Science* series. Its focus on the topic of life science is timely and needed. Since 2005, Page Keeley and her coauthors have developed 170 successful probes in physical, life, and Earth and space science. Until now, they had not devoted an entire book to life science probes. The probes in this book capture the many dimensions of biology, not just the animal and plant side of the field. Included in the mix of 25 probes are those that focus on the cellular and ecological, as well as the genetics and zoological, domains of life science. Further, the biological ideas that are presented in this book are aligned with the real-life phenomena that students encounter in and outside their classrooms.

As a community of science educators, we are lucky to have Page Keeley and her commitment to the task of capturing student thinking. Her passion in this area has made an important contribution to the field of science education, with both teachers and students gaining tremendously.

Julie Luft
Arizona State University
NSTA, Division Director, Research in Science Education

References

National Research Council (NRC). 1996. *National science education standards.* Washington, DC: National Academies Press.

American Association for the Advancement of Science (AAAS). 1993/2009. *Benchmarks for science literacy.* Washington, DC: AAAS. Online at *www.project2061.org/publications/bsl/online*

Preface

Series Overview: *Uncovering Student Ideas in Science*

This book is one of a "series of series" of unique K–12 formative assessment resources. The first book in the first series was *Uncovering Student Ideas in Science: 25 Formative Assessment Probes* (Keeley, Eberle, and Farrin 2005). Subsequently, NSTA published three more books of K–12 formative assessment probes.* The four books in the *Uncovering Student Ideas in Science* series cover a wide range of probes in the life, Earth, space, and physical sciences as well as the nature of science and "unifying themes" such as systems and models.

In 2010, NSTA published the first volume in a new physical science series—*Uncovering Student Ideas in Physical Science, Vol. 1: 45 New Force and Motion Assessment Probes* (Keeley and Harrington 2010). The four additional books planned for this series will cover electricity and magnetism, energy, light and sound, and matter and energy.

The present book, *Uncovering Student Ideas in Life Science, Vol. 1: 25 New Formative Assessment Probes,* addresses K–12 ideas on the topics of life and its diversity; structure and function; life processes; ecosystems and change; reproduction, life cycles, and heredity; and human biology. This volume will be followed by a second life science volume, and possibly a third. It seems there is never a shortage of key science ideas commonly held by students that can be elicited by formative assessment probes. As long as there are interesting and important science ideas to examine—and teachers who want to learn more about how their students' think about these ideas—this series will continue to provide new and informative volumes on additional topics such as astronomy, the nature of science, and Earth science.

Although some diagnostic assessment tools for science teachers do exist, mostly in the physical sciences, there are few diagnostic tools that can be used in grades K–12 in all of the sciences. The importance of identifying and analyzing students' preconceptions (diagnostic assessment) and then using the data on students' preconceptions to inform teaching and learning (formative assessment) was recognized during the 1990s in publications such as *How People Learn: Brain, Mind, Experience, and School* (Bransford, Brown, and Cocking 2000). It became clear to science teachers and educators that there was a need for student- and teacher-friendly probing questions to access students' thinking and prior knowledge in science and to promote learning by involving students in the examination of their own and their peers' ideas. This need led to the development of this popular series.

With the addition of the present book, the *Uncovering* series consists of a total collection of 170 formative assessment probes. These probes are used by thousands of K–12 classroom teachers, university professors, informal educators, professional developers, instructional coaches and mentors, and even parents, and they consistently appear on the NSTA Press list of bestsellers. The series has become a valuable resource not only for improving student learning but also for deepening teachers' understanding of science content and of pedagogical content knowledge (PCK). (PCK is the specialized knowledge of teaching that

* See References, pp. xvii–xviii: Keeley, Eberle, and Tugel 2007; Keeley, Eberle, and Dorsey 2008; Keeley and Tugel 2009.

teachers—in our case, science teachers—have, including knowledge of the most effective ways of presenting a topic to make it comprehensible to learners. To teach all students according to today's standards, teachers need to have a deep and flexible understanding of subject matter so that they can help students create useful cognitive maps, connect one idea to another, and address misconceptions [Shulman 1992].)

Each book in the original *Uncovering* series begins with a unique introduction to some aspect of formative assessment.

- Introduction to Vol. 1: Offers an overview of formative assessment: what it is and how it differs from summative assessment. Provides background on probes as specific types of formative assessments and how they are developed.
- Introduction to Vol. 2: Describes the link between formative assessment and instruction and suggests ways to embed the probes into your teaching.
- Introduction to Vol. 3: Describes how you can use the probes and student work for teacher learning—either individually or through professional learning communities—to (a) deepen your understanding of students' ideas and their implications for instruction, (b) learn new science content, or (c) even uncover a deeply rooted misconception you might have.
- Introduction to Vol. 4: Describes the link between formative and summative assessment, including reasons why an investment in formative assessment before and throughout instruction can improve students' performance on the summative end.
- *Uncovering Student Ideas in Physical Science, Vol. 1:* The introduction to this book provides an overview of the teaching difficulties and research about students' ideas related to force and motion concepts and strategies for effectively addressing students' ideas.

Collectively, the introductory chapters of the five previous volumes, in addition to the introduction in this volume (p. 1), will expand your assessment literacy and understanding of effective teaching and learning and will deepen your understanding of students' thinking about the important content of science.

About the Probes

A *probe* is a specific type of question designed to reveal more than just an answer. A probe uncovers significant data about students' thinking—for example, about their scientifically correct ideas, misconceptions, partially formed ideas, and the types of reasoning and connections they use to make sense of phenomena or concepts. The probes in the *Uncovering* series are considered two-tiered assessments. The first tier is a forced-choice response where students select from a list that includes the answer and several distracters. The second tier requires students to explain their thinking by explaining why they selected a particular answer.

This two-tiered approach is essential to formative assessment because students may answer a selected response question correctly but have major flaws in their reasoning for choosing that response. Conversely, they may select an incorrect response but show sophisticated reasoning with pieces of correct knowledge that have not been connected in a coherent way. In both cases, all the information is useful to the teacher when determining the learning experiences that he or she will need to provide to move students from where they are in their current thinking to where they need to be to achieve scientific understanding.

The probes in all of the books in the *Uncovering* series were developed using a process that links key ideas in science to children's commonly held ideas as revealed in research about how children learn. This formative assessment probe development process, supported with funding from the National Science

Foundation, is explained in *Science Curriculum Topic Study: Bridging the Gap Between Standards and Practice* (Keeley 2005). Tools, templates, and detailed instructions for facilitating the assessment probe development process are described in *A Leader's Guide to Science Curriculum Topic Study* (Mundry, Keeley, and Landel 2009). Both of these publications are available through NSTA Press (*www.nsta.org/store*) and are highly recommended if you would like to increase your knowledge about the link between science standards, research on learning (including misconceptions), and classroom practice.

About the Teacher Notes

The Teacher Notes that follow each probe provide important information related to the content of the probe, the content's connection to science standards, grade-level considerations, the underlying research on students' commonly held ideas, and suggestions for instruction, including suggestions for addressing the ideas your students have related to the probe and for informing your own professional knowledge. The Teacher Notes that accompany each of the probes are made up of these 11 sections:

1. Purpose

This section begins by describing the general concept elicited by the probe and then states the specific ideas that are targeted by the probe. Before you choose a probe, it is important that you read this section to make sure that the probe is designed to provide the information about students' ideas that you need to inform your teaching.

2. Related Concepts

Each probe is designed to target one or more related concepts that often cut across grade spans, such as the concept of a cell. These concepts are also included on the concept matrix charts on pages 8 and 84. A single concept may be addressed by more than one probe, as indicated on the concept matrix. You may find it useful to use a cluster of probes to target a concept or specific ideas within a concept. For example, there are four probes related to the concept of photosynthesis.

3. Explanation

A brief scientific explanation accompanies each probe and provides clarification of the scientific content that underlies the probe. The explanations are designed to help you identify what the "best" or most scientifically acceptable answers are (sometimes there is an "it depends" answer) and to clarify any misunderstandings you might have about the content. Great care was taken not to oversimplify the content explanations for teachers with biology backgrounds. Conversely, explanations were carefully worded to provide the information that novice life science teachers with little or no formal backgrounds in biology would need to understand the content.

If you need additional background information on the content, please refer to the references and the related NSTA resources that appear at the end of each Teacher Notes.

4. Curricular and Instructional Considerations

Unlike summative assessments, the probes in this book do not target a single grade level. Rather, they provide insights into the knowledge and thinking your students may have regarding a topic as they developmentally progress or move from one grade span (i.e., elementary, middle, or high school) to the next. Some of the probes can be used in K–2, 3–5, 6–8, and 9–12; others address a single grade span, such as 9–12. Teachers from two different grade spans might decide to administer the same probe and then come together to discuss their findings. To do this, it is helpful to know what students typically experience and are expected to know at a given grade span about the ideas elicited by the probe. Because the probes do not typically identify one specific grade level for use, you are encouraged to read the Curricular and Instructional Considerations sections and then decide if your students have sufficient readiness to respond to the probe and whether the information you will get from the probe is likely to be useful in your particular context.

5. Administering the Probe

This section provides suggestions for administering the probe to students, including what the appropriate grade levels are and how various modifications can make the probe more accessible for certain students. For example, the notes might recommend eliminating some of the choices from the list of possible answers for younger students, who may not be familiar with particular words or examples. This section also tells teachers when it is a good idea to establish the context for a probe by showing students various items or props referred to in the probe (e.g., showing students an unopened packet of cucumber seeds or a picture of a seed packet when using the "Cucumber Seeds" probe on p. 9). This section often suggests ways to elicit probe responses when students are working in collaborative groups or engaging in a whole-class discussion.

Most of the probes are not grade-level-specific in the way summative assessments are designed. The suggested grade level is intended to be a guide and depends on your use of the probe and your students' readiness. You might think about these questions:

- Do you want to know what ideas the national science standards expect students at your grade level to learn?
- Are you interested in how preconceived ideas develop and how they are apt to change across multiple grade levels, whether or not they are formally taught?
- Are you interested in whether your students understand previous-grade-level science concepts before you introduce higher-level concepts?
- Do you have students who are ready for advanced concepts that exceed their grade-level expectations?

Weigh the suggested grade levels in the Administering the Probe section against the knowledge you have of your own students and your school's curriculum.

6. Related Ideas in the National Science Standards

This section lists the learning goals stated in the two major documents generally considered to be the national science standards—the most recent version of *Benchmarks for Science Literacy* (AAAS 2009) and *National Science Education Standards* (NRC 1996). Because these are the primary documents on which almost all current state science standards are based, it is important to look at the related learning goals in these documents. In future publications in this series, we plan to include the new Next Generation Science Standards, now under development by the National Research Council.

The learning goals in this section (which are quoted directly from the two primary documents) are not intended to align directly to the probe. Rather they appear in a probe because they are closely related to it in some way. For example, some targeted probe ideas, such as the tropism-related ideas in "Pumpkin Seeds" and "Rocky Soil," although not explicitly stated as learning goals in the national science standards, are clearly related to concepts regarding specific ideas about the behavior of organisms. The topic of tropisms is not included in the standards for the purpose of learning the names and types of different plant responses.

Rather, it serves as a context for important learning goals related to behavioral responses to changes in an organism's environment. Therefore, the two probes that address plants' gravitropic or thigmotropic responses to environmental stimuli (#12 and #13) clearly relate to the National Science Education Standards middle and high school ideas related to the behavioral characteristics of organisms.

Whenever the ideas elicited by a probe *do* appear to be a strong match (aligned) with a national science standard's learning goal, these matches are indicated by a star symbol (★). You may find this information helpful when using probes with lessons and instructional materials that are strongly aligned to a similar goal in your local standards, curriculum materials, and specific grade level.

Sometimes you will notice that an elementary learning goal is included with probes that have been designated for middle and high school use. This is because it is often useful to see the related idea that the probe builds on from a previous grade span. Likewise, a high school learning goal is sometimes included with a probe for grades K–8. This is because it is useful to consider the *next level* of sophistication that students will encounter in their spiraled learning.

7. Related Research

Each probe is informed by related research, where available, or results from the field tests that were conducted during development of the probes. Three sources of comprehensive research summaries readily available to educators were used to describe the commonly held ideas that students have about the life science concepts addressed in this book: Chapter 15 in *Benchmarks for Science Literacy* (AAAS 1993, 2009); the research summaries in the two volumes of the *Atlas of Science Literacy* (AAAS 2001, 2007); and *Making Sense of Secondary Science: Research Into Children's Ideas* (Driver et al. 1994). Available research in the specific field of biology education was also used.

It should be noted that although many of the research articles cited in this section of a probe describe studies that were conducted in past decades and involved children in other countries as well as the United States, most of the results of these studies are considered timeless and universal. Whether students develop their ideas in the United States or other countries, research indicates that many of their preconceptions about living phenomena and life science concepts are similar—regardless of geographic boundaries and societal and cultural influences. Hence the descriptions from the research can help you anticipate the kinds of thinking your students are likely to reveal when they respond to a probe and the factors that may have influenced their thinking. As you use the probes, you are encouraged to explore current and readily available published research. The Curriculum Topic Study website at *www.curriculumtopicstudy.org* has a searchable database where you can access additional current research articles on learning.

8. Suggestions for Instruction and Assessment

A probe is simply diagnostic, not formative, unless you use the information acquired from examining your students' thinking to inform your instruction. After analyzing your students' responses, you will then want to take two important steps: deciding on individual student interventions that differentiate instruction according to the various ideas students may hold and formulating an instructional path that can be used to guide whole-class instruction. You may find that additional probing assessments are needed. Suggestions in this section have been gathered from the wisdom of teachers, the knowledge base on effective science teaching, research on various strategies for learning, and the author's and her colleagues' extensive experience working with students and teachers.

In this section, you will not find detailed descriptions of instructional activities but rather brief suggestions for planning or modifying your curriculum or instruction to help your students cross the bridge that begins with their own preconceptions and leads them toward a scientific understanding. Your role may be as simple as making sure that your students are not limited by the context in which they learned a particular concept. For example, the probe "Eggs" on page 117 may point out the limitations of context when teaching about life cycles or how animals are born. Some students may believe that only organisms that hatch from eggs that develop in the external environment begin their life cycle as an egg. They may not recognize that animals that give birth to live young also started life as an egg. You should also be aware during your teaching that the use of an everyday word such as *animal* can create confusion among students when the teacher is using such a word according to its different, biological meaning—for example, some students may not realize that, in biology, humans are considered animals (see "No Animals Allowed," p. 21).

Learning is a very complex process and no single suggestion will help all students learn. But formative assessment encourages you to think carefully about how to move your students conceptually from where they are initially in their learning to where they need to be. All the while, you are monitoring their progress during the course of instruction and learning. As you become more familiar with the ideas your students have and the many factors that may have contributed to their misunderstandings, you will identify additional strategies to teach for conceptual change.

9. Related NSTA Science Store Publications, NSTA Journal Articles, NSTA Learning Center Resources (SciGuides, SciPacks, and Science Objects)

NSTA Press books and NSTA's journals, SciGuides, SciPacks, and Science Objects are increasingly targeting the ideas students bring to their learning. In this section, I provide suggestions for additional readings and other curricular materials that complement or extend the use of the individual probes and the background information that accompanies them. For example, NSTA Press's *Hard-to-Teach Biology Concepts: A Framework to Deepen Student Understanding* (Koba and Tweed 2009) can clarify biology concepts that teachers and students struggle with. Journal articles from NSTA's elementary, middle school, and high school member journals can provide additional insight into students' misconceptions about food chains and food webs or present effective instructional strategies or activities for deepening student understanding of the ideas targeted by a probe. The NSTA Science Object called *Cell Structure and Function: Cells—The Basis of Life* can be used to deepen a novice life science teacher's content knowledge about cells.

10. Related Curriculum Topic Study Guides

NSTA is the copublisher of *Science Curriculum Topic Study: Bridging the Gap Between Standards and Practice* (Keeley 2005). This book was developed as a professional development resource for teachers with funding from the National Science Foundation and is available through NSTA Press. It provides a set of 147 curriculum topic study (CTS) guides that can be used to learn more about a science topic's content, examine instructional implications, identify specific learning goals and scientific ideas, examine the research on student learning, consider connections to other topics, examine the coherency of ideas that build over time, and link understandings to state and district standards. The CTS guides use national standards and research in a systematic process that deepens teachers' understanding of the topics they teach.

The specific CTS guides that informed the development of the probes in this book appear in boxes before the References in the Teacher Notes. Teachers who wish to delve deeper into the standards and research-based findings that were used to develop the probes may wish to use the guides for further information. In addition, the CTS website—*www.curriculumtopicstudy.org*—provides a database of supplemental resources linked to each topic covered in the book. These supplemental resources include new research on students' ideas, videos that address the topic, and professional development materials.

11. References

References conclude each of the Teacher Notes. The references provide the complete citations for publications and other materials cited in the Teacher Notes. Consider reading some of these articles or books to really dig down into a subject that interests you.

Formative Assessment Reminder

Now that you have some background on the older series, this new series, the probes, and the Teacher Notes, let's not forget the formative purpose of these probes. *Remember—a probe is not formative unless you use the information from the probe to modify, adapt, or change your instruction so that students have opportunities to learn certain important life science concepts.*

As a companion to this book and to the others in the *Uncovering* series, NSTA has copublished *Science Formative Assessment: 75 Practical Strategies for Linking Assessment, Instruction, and Learning* (Keeley 2008). This book contains a variety of strategies to use with the probes to facilitate elicitation of student thinking, support metacognition, spark inquiry, encourage discussion, monitor progress toward conceptual change, encourage feedback, and promote self-assessment and reflection. By using the probes and their supporting formative assessment classroom techniques (FACTs), you will create learning environments that acknowledge that all students' ideas are important. Taking the time to make students' ideas visible will help you discover and use new knowledge about teaching and learning and engage your students in the process of conceptual change.

If you have stories to share about your use of the probes or suggestions for future probe topics, please feel free to e-mail me at *pagekeeley@gmail.com*. In addition, if you are interested in planning a workshop or other professional development sessions using the *Uncovering* series and its formative assessment classroom techniques, check out NSTA Press's Authors Speak site—*www.nsta.org/publications/press/authorsspeak*—through which you can arrange to bring an NSTA Press author, such as the author of this series, to your school or district for professional development. Following an introduction to using formative assessment and the assessment probes, many school districts throughout the country, both large and small, are recognizing the importance of using embedded professional development, such as book studies, to build and sustain a collaborative culture of formative assessment in their schools.

References

American Association for the Advancement of Science (AAAS). 1993. *Benchmarks for science literacy.* New York: Oxford University Press.

American Association for the Advancement of Science (AAAS). 2001. *Atlas of science literacy.* Vol. 1. Washington, DC: AAAS.

American Association for the Advancement of Science (AAAS). 2007. *Atlas of science literacy.* Vol. 2. Washington, DC: AAAS.

Preface

American Association for the Advancement of Science (AAAS). 2009. Benchmarks for science literacy online. *www.project2061.org/publications/bsl/online*

Bransford, J., A. Brown, and R. Cocking. 2000. *How people learn: Brain, mind, experience, and school.* Washington, DC: National Academies Press.

Driver, R., A. Squires, P. Rushworth, and V. Wood-Robinson. 1994. *Making sense of secondary science: Research into children's ideas.* London: Routledge-Farmer.

Keeley, P. 2005. *Science curriculum topic study: Bridging the gap between standards and practice.* Thousand Oaks, CA: Corwin Press and Arlington, VA: NSTA Press.

Keeley, P. 2008. *Science formative assessment: 75 practical strategies for linking assessment, instruction, and learning.* Thousand Oaks, CA: Corwin Press and Arlington, VA: NSTA Press.

Keeley, P., F. Eberle, and C. Dorsey. 2008. *Uncovering student ideas in science, vol. 3: Another 25 formative assessment probes.* Arlington, VA: NSTA Press.

Keeley, P., F. Eberle, and L. Farrin. 2005. *Uncovering student ideas in science, vol. 1: 25 formative assessment probes.* Arlington, VA: NSTA Press.

Keeley, P., F. Eberle, and J. Tugel. 2007. *Uncovering student ideas in science, vol. 2: 25 more formative assessment probes.* Arlington, VA: NSTA Press.

Keeley, P., and R. Harrington. 2010. *Uncovering student ideas in physical science, vol. 1: 45 new force and motion assessment probes.* Arlington, VA: NSTA Press.

Keeley, P., and J. Tugel. 2009. *Uncovering student ideas in science, vol. 4: 25 new formative assessment probes.* Arlington, VA: NSTA Press.

Koba, S., with A. Tweed. 2009. *Hard-to-teach biology concepts: A framework to deepen student understanding.* Arlington, VA: NSTA Press.

Mundry, S., P. Keeley, and C. J. Landel. 2009. *A leader's guide to science curriculum topic study.* Thousand Oaks, CA: Corwin Press.

National Research Council (NRC). 1996. *National science education standards.* Washington, DC: National Academies Press.

Shulman, L. 1992. Ways of seeing, ways of knowing, ways of teaching, ways of learning about teaching. *Journal of Curriculum Studies* 28: 393–396.

Acknowledgments

I would like to thank the teachers and science coordinators I have worked with for their willingness to field-test probes, share student data, and contribute ideas for assessment probe development. I would especially like to acknowledge Amy Neiman, a teacher at La Center High School in La Center, Washington. Amy's use of the "Let's Keep Thinking!" sticky bars technique was the inspiration for the vignette in the introduction of this book. Thanks, Amy, for sharing such a creative way to use this technique to monitor changes in students' thinking throughout instruction! Also, my special thanks go to Beth Chagrasulis at Bridgton Academy in North Bridgton, Maine. Beth is a talented high school biology teacher who never hesitates to take the time to give me feedback on the probes and a window into how they can be used in creative ways.

I would also like to thank my colleagues at the Maine Mathematics and Science Alliance (MMSA) (*www.mmsa.org*) who support me in this work as well as my professional development colleagues, the Math-Science Partnership directors, and university partners throughout the United States whom I have had the pleasure of sharing the *Uncovering Student Ideas* work with in various projects. I also thank the book's four reviewers for providing useful feedback to improve the original manuscript: Susan Brady, director, Center for Science and Engineering, University of California; A. Daniel Johnson, senior lecturer and Kirby Faculty Fellow, Department of Biology, Wake Forest University; Richard Konicek-Moran, educator, biologist, and professor emeritus, University of Massachusetts-Amherst; and Luke Sandro, biology teacher, Springboro (Ohio) High School.

I especially thank Dr. Julie Luft for taking the time to write the foreword for this first volume in the life science series. Julie is an inspiring researcher and educator who truly walks the talk when it comes to building a bridge between research and practice. And deep gratitude goes to all my friends at NSTA Press—it is a privilege to work with you!

About the Author

Page Keeley is the senior science program director at the Maine Mathematics and Science Alliance (MMSA) where she has worked since 1996. She directs projects in the areas of leadership, professional development, linking standards and research on learning, formative assessment, and mentoring and coaching, and she consults with school districts and organizations nationally. She was the principal investigator on three National Science Foundation grants: the Northern New England Co-Mentoring Network; Curriculum Topic Study: A Systematic Approach to Utilizing National Standards and Cognitive Research; and PRISMS: Phenomena and Representations for Instruction of Science in Middle School. She is the author of 10 books (including this one): four books in the *Curriculum Topic Study* series (Corwin Press); five volumes in the *Uncovering Student Ideas in Science* series (NSTA Press); and *Science Formative Assessment: 75 Practical Strategies for Linking Assessment, Instruction, and Learning* (Corwin Press and NSTA Press).

Most recently she has been consulting with school districts, Math-Science Partnership projects, and organizations throughout the United States on building teachers' capacity to use diagnostic and formative assessment. She is frequently invited to speak at national conferences, including the annual conference of the National Science Teachers Association. She led the People to People Citizen Ambassador Program's Science Education delegation to South Africa in 2009 and to China in 2010.

Page taught middle and high school science for 15 years; in her classroom she used formative assessment strategies and probes long before there was a name attached to them. Many of the strategies in her books come from her experiences as a science teacher. During her time as a classroom teacher, Page was an active teacher leader at the state and national level. She received the Presidential Award for Excellence in Secondary Science Teaching in 1992 and a Milken National Distinguished Educator Award in 1993. She was the AT&T Maine Governor's Fellow for Technology in 1994, has served as an adjunct instructor at the University of Maine, is a Cohort 1 Fellow in the National Academy for Science and Mathematics Education Leadership, and serves on several national advisory boards.

Prior to teaching, she was a research assistant in immunology at the Jackson Laboratory of Mammalian Genetics in Bar Harbor, Maine. She received her BS in life sciences from the University of New Hampshire and her MEd in secondary science education from the University of Maine. Page was elected the 63rd president of the National Science Teachers Association for the 2008–2009 term. In 2009 she received the National Staff Development Council's Susan Loucks-Horsley Award for her contributions to science education leadership and professional development.

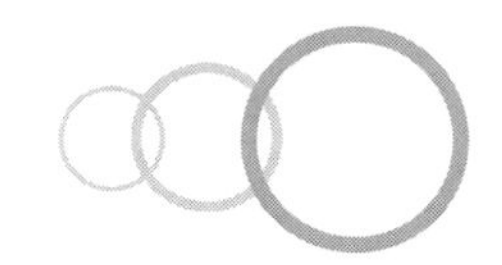

Introduction

> "Critical to effective science instruction that results in robust understanding is a teacher who monitors what students think and probes that thinking, posing careful questions to challenge students' ideas and using that information to structure future questions, activities, and assessments."
>
> —Cox-Petersen and Olson (2002)

Mrs. Jimenez browsed through the book *Uncovering Student Ideas in Life Science* (actually, this book!) in search of a probe to use before planning her sequence of instruction on the life processes of plants. She decided to use the friendly-talk* probe "Light and Dark" (p. 63) to find out what preconceptions her

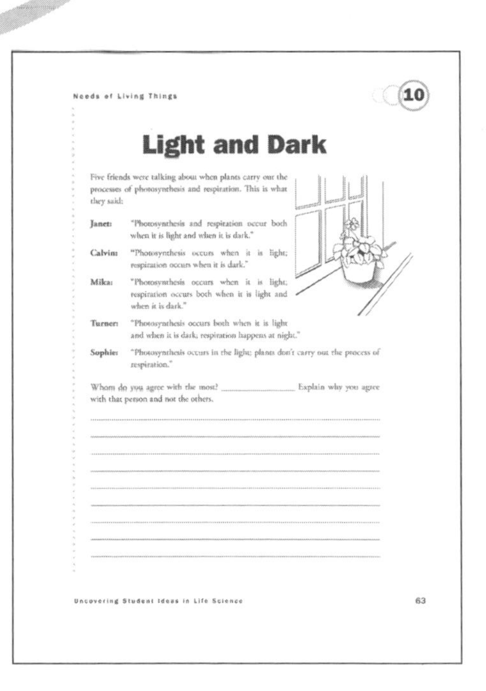

Needs of Living Things

10

Light and Dark

Five friends were talking about when plants carry out the processes of photosynthesis and respiration. This is what they said:

Janet: "Photosynthesis and respiration occur both when it is light and when it is dark."

Calvin: "Photosynthesis occurs when it is light; respiration occurs when it is dark."

Mika: "Photosynthesis occurs when it is light; respiration occurs both when it is light and when it is dark."

Turner: "Photosynthesis occurs both when it is light and when it is dark; respiration happens at night."

Sophie: "Photosynthesis occurs in the light; plants don't carry out the process of respiration."

Whom do you agree with the most? ______________ Explain why you agree with that person and not the others.

Uncovering Student Ideas in Life Science 63

---

* A friendly-talk probe is a question set in the context of a group of friends talking about their ideas. The answer choices use the names of the friends instead of, for example, A, B, C, D. Students often find this form of question much more engaging than a standard multiple-choice question. The friendly-talk probes in this book are #3, 5, 9, 10, 11, 12, 13, 15, 16, 17, 18, 20, 21, and 23.

high school biology students had about photosynthesis and plant respiration.

Prior to selecting the probe, Mrs. Jimenez had examined the middle school curriculum in her school district to identify the key ideas related to photosynthesis and cellular respiration that her students had been introduced to in the seventh grade. Then she read the Teacher Notes for "Light and Dark" and learned that students sometimes believe that photosynthesis and respiration are opposite processes. Because animal and plant processes are often taught separately, some students may form the erroneous "opposites" belief that respiration is an animal process and photosynthesis is a plant process. Even students who recognize that plants respire continue to use the "rule of opposites" in believing that photosynthesis happens only in the daytime and respiration happens only at night (i.e., that with plants photosynthesis takes place in the light and respiration occurs in the dark). She also read that the terms *light reaction* and *dark reaction*—terms commonly used by science teachers—may also contribute to students' misconceptions. She decided that "Light and Dark" would be a good probe to use for uncovering some of these areas of confusion. She then planned her instruction accordingly so that students would learn to distinguish between the two processes by comparing their similarities and differences.

Mrs. Jimenez has used formative assessment probes and FACTs (formative assessment classroom techniques) described in another one of her resources, *Science Formative Assessment: 75 Practical Strategies for Linking Assessment, Instruction, and Learning* (Keeley 2008), quite a bit over the last few years. She really likes the way the techniques engage all students and encourage them to be aware of not only their own ideas and ways of thinking but also those of their peers. Since using the probes and various FACTs, she has seen her students become more confident in their thinking and more open to sharing their ideas, regardless of whether their ideas are right or wrong. She feels her classroom has evolved into a "thinking classroom"—a place where her students continually revisit their ideas and work together to discard explanations whose implausibility becomes obvious to them in light of new scientific understandings.

Mrs. Jimenez likes to visually display the variety of student responses to a probe so that all students can see what others in the class think. She doesn't have a set of electronic personal response units (also called clickers) that can be used to display student responses, but that is not a problem. She uses low-tech sticky notes to build what is called a Sticky Bars Chart (Keeley 2008). She gives each student a probe handout and a sticky note. Each student writes his or her response to the probe on the sticky note (e.g., a, b, or c). The notes are then collected and posted as a bar graph, showing the numbers of responses for each answer choice. The students really enjoy using this technique because it allows them to share their answers anonymously, see the variety of responses held by class members, and realize that not everyone knows the right answer initially.

Mrs. Jimenez recently attended a science formative assessment workshop. The workshop presenter shared an interesting adaptation of the sticky bars technique, developed by a high school teacher in Washington State, using the "Giant Sequoia Tree" probe from *Uncovering Student Ideas in Science, Vol. 2* (Keeley, Eberle, and Tugel 2007). Instead of using sticky bars **only** at the beginning of the unit, the teacher uses them at the beginning, midway, and at the end so that students can see how the class collectively modified their ideas as they progressed through the unit.

Mrs. Jimenez was intrigued by the idea of revisiting the probe after students had gained more information and experiences in the lab, so she decided to try that adaption of the technique. She set up a chart titled "Light and Dark—Let's Keep Thinking!" At the bottom of the chart, she wrote the students' names from the friendly-talk probe "Light and Dark" (Janet, Calvin, Mika, Turner, and Sophie). On Monday morning, Mrs. Jimenez announced that the topic of their new unit was photosynthesis and respiration. She shared her learning goals and the essential question: "How are the processes of photosynthesis and respiration alike and different?" She explained that before they began reading about, discussing, and investigating the topic through various reading assignments, labs, and computer activities, they were going to find out what their ideas were *right now* about photosynthesis and respiration.

She gave each student a copy of the probe handout "Light and Dark" (p. 63 in this book) and a green sticky note. She asked her students to read the probe and then write down on the green sticky note the response that best matched their current thinking (in this case, the response of Janet, Calvin, Mika, Turner, or Sophie). She collected the sticky notes and with the help of a couple of students, made a bar graph on the chart she had previously prepared (see Figure 1). In the upper right-hand corner, she posted a sticky note with the number 1. The number would remind the class that they were on their first round of responding to the probe.

Of the 20 students in the class, only three students chose the "best" response (Mika's): "Photosynthesis occurs when it is light; respiration occurs both when it is light and when it is dark." Of course, at this time the students wanted Mrs. Jimenez to tell them the right answer! Some students were sure that the right answer must be Calvin's because the majority of students selected that response. (Calvin's response was, "Photosynthesis occurs when it is light; respiration occurs when it is dark.") Mrs. Jimenez remembered how important it is

Figure 1. Sticky Bars Chart: Initial Class Responses

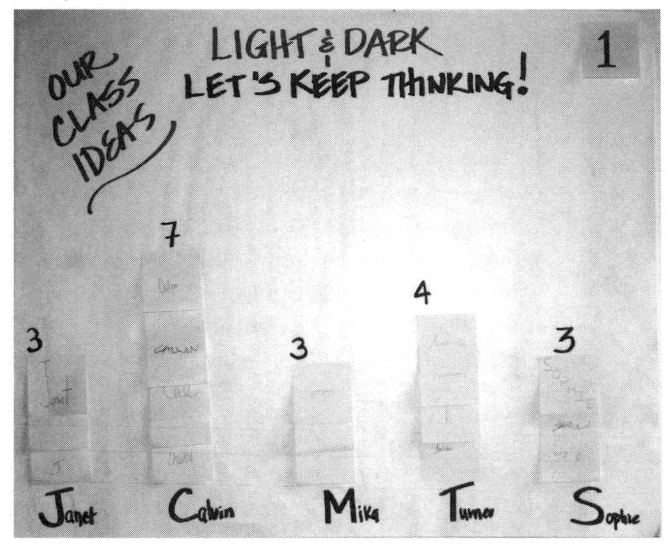

for the students to "hang out in ambiguity" for a while and keep thinking. She told them that over the next several days they would be learning more about the processes of photosynthesis and respiration and would revisit the probe to see if and how their ideas had changed. In the meantime, she asked students to write an explanation for their answers on the probe handouts. She collected these and after school read the students' explanations.

She was surprised to find several misconceptions that might be traced back to conceptual misunderstandings some students might have developed during middle school that went unnoticed by their teachers. For example, a large number of students selected Calvin's response and explained that photosynthesis and respiration are opposite processes—so, if one must occur in the light, then the other must occur in the dark *because* they are opposite processes. Mrs. Jimenez wondered if students constructed this rule because they know that plants take in carbon dioxide and give off oxygen during photosynthesis and take in oxygen and give off carbon dioxide during respiration. Perhaps they thought that if one opposite rule applies then other opposite rules apply as well.

She also noted that three of her students chose Sophie's response, believing that only animals carry out the process of respiration. This was not really a surprise to Mrs. Jimenez because she had read the Teacher Notes before distributing the probe handouts, which had alerted her to this common misconception. After giving a great deal of thought to the students' responses, Mrs. Jimenez began to plan her unit so that her instruction would address her students' ideas.

Fast forward five days later. After a series of online activities, labs, readings, and discussions, Mrs. Jimenez announced that they would revisit the "Light and Dark" probe. She returned the original handouts to the students and asked them to decide whether they still agreed with their initial answers or whether their ideas had changed. She pointed to the class's Sticky Bar Chart and explained that today she would give each student a new sticky note of a different color (blue) to record the responses that best matched their thinking now.

She said that if their answers today were the same as their original answers, they should elaborate on their initial explanations or explain why their thinking had not changed. If they changed their answers, they should provide new, revised explanations. She passed out the blue sticky notes and then collected them after everyone had responded. She created a double bar graph, placing the blue bars next to the green bars (see Figure 2, where the darker bars are the blue bars). She placed a blue sticky note, labeled #2 under the #1 note to indicate that the blue stickies represented the second time they answered the probe.

The students had a lively discussion about the differences in their responses to the probe before they started their unit and their responses midway into the unit. Several of the students mentioned the "light reactions" and "dark reactions" of photosynthesis described in their text, even though Mrs. Jimenez had been careful not to use those terms (which describe two parts of the photosynthetic reaction—one that is light dependent, one that is not light dependent). She made a note that she needed to further address the "dark reaction" by referring to it as the Calvin cycle, explaining how light is not necessary for this phase, yet making sure that students understand that the photosynthetic process as a whole requires light.

After several spirited arguments, the class as a whole decided they could discard Sophie's answer, and all but one student were ready to discard Calvin's answer. Mrs. Jimenez made a note that she needed to reinforce the idea that respiration occurs continuously in plants, just as it does in animals. Perhaps if she tied this idea more firmly to the continuous need for energy, her students would be more apt to accept it.

Fast forward to the next week, the end of the unit. Mrs. Jimenez concluded her unit by

Figure 2. Sticky Bars Chart: Revisited After Five Days

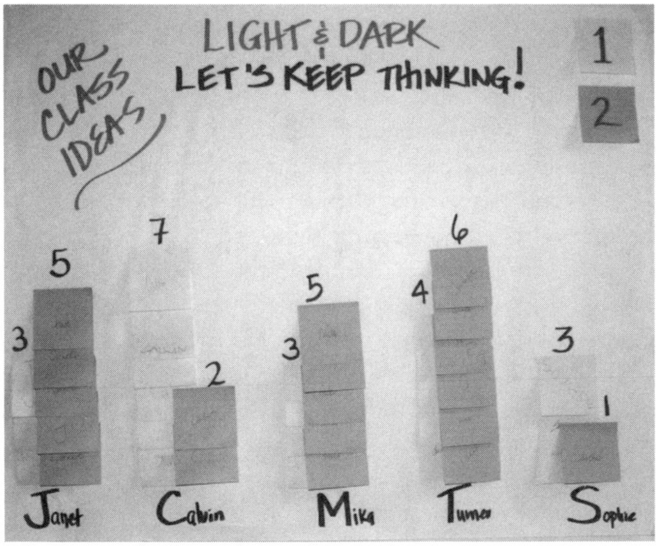

asking the students to reflect on what they had learned. She pointed to the "Light and Dark—Let's Keep Thinking!" chart and told the class they would have one last chance to reflect on their previous ideas and either change or keep their previous ideas. She again returned their original handouts to them and this time gave each student a pink sticky note, asking them to respond to the probe one more time. She told them to reflect back on all that they had learned and experienced and use their present understandings and experiences as evidence to support their final answers and explanations. She collected the sticky notes, added a third, pink bar to the bar graph for the third and final response, and a number 3 in the upper right-hand corner of the chart (see Figure 3, where the darkest bars are the pink bars).

Figure 3. Sticky Bars Chart: Revisited at End of Unit

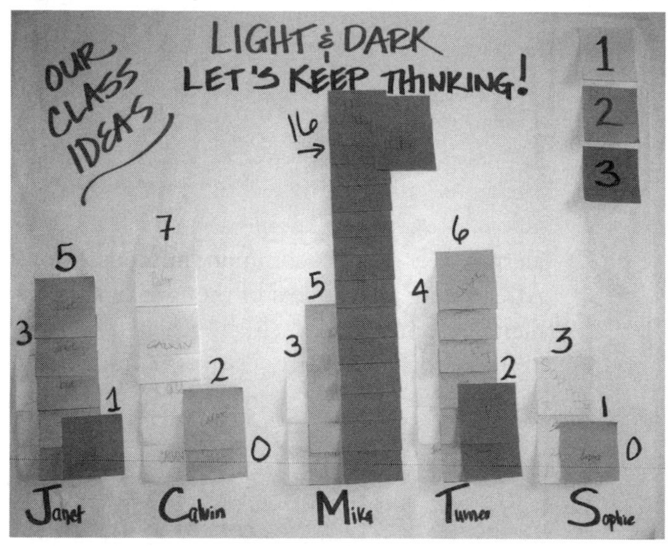

Wow! She and the class noticed how many students now selected Mika's answer (the best answer), even though there were still a few students who could not give up their previous conceptions. Mrs. Jimenez asked the students who had chosen Mika as their answer to share their explanations, including why they may have changed their previous answers. They supported their explanations with evidence from the lessons they had experienced. Hearing them, one of the students who had chosen Turner's answer said, "OK, I get it now. I'm willing to accept Mika's idea."

Two of the students were still not convinced. Mrs. Jimenez made note of their reluctance to change their answers, knowing that some students, even after instruction, still cling tenaciously to their original ideas. She also knew that *merely correcting them and moving on wouldn't change their thinking.* She planned to bring up these concepts again at an appropriate time. Perhaps then those students would finally be ready to accept the scientific idea. In the meantime Mrs. Jimenez was pleased to see that her instruction resulted in conceptual change for a significant number of students, and, even more important, the students recognized the change as well!

Sometimes the learning bridge is short, sometimes it's long, but the goal is to move as many students as possible to the other side in the limited school time we have available. Assessment probes are a tremendous tool for constructing that bridge. However, what about the students who don't make it to the other side? This vignette from Mrs. Jimenez's class reinforces the fact that teaching difficult ideas is not a one-shot deal. Difficult ideas need to be revisited in different contexts or with different probes whenever the opportunity arises. After all, that's what good teaching is about—moving students from where they are to where they need to be in order to have a scientific understanding, while recognizing that everyone doesn't move at the same pace.

This vignette points out the importance of using formative assessment probes to uncover students' existing ideas and then to use that information to plan targeted instruction. In addition, probes and FACTs can be used to continuously monitor students' thinking and their progress toward achieving conceptual change. As you use the life science probes in this volume, think about the essential ideas you want to target in your curriculum. The probes can help you discover barriers to learning that often remain hidden if they are not deliberately uncovered by the teacher. However, merely uncovering students' alternative ideas is not enough. Students must be confronted with them and guided through a sense-making process that gives them the new evidence they need to willingly give up or revise their alternative ideas when they realize that these ideas no longer hold explanatory power. With this book, you have the tools to uncover a variety of alternative ideas (as well as correct ones) that your students may have about life science concepts. However, the most important step is deciding on the actions you will take as a teacher to address your students' ideas. And that is the essence of good science teaching that supports conceptual change!

References

Cox-Petersen, A., and J. Olson. 2002. Assessing student learning. In *Learning science and the science of learning,* ed. R. W. Bybee, 105–118. Arlington, VA: NSTA Press.

Keeley, P. 2008. *Science formative assessment: 75 practical strategies for linking assessment, instruction, and learning.* Thousand Oaks, CA: Corwin Press and Arlington, VA: NSTA Press.

Keeley, P., F. Eberle, and J. Tugel. 2007. *Uncovering student ideas in science, vol. 2: 25 more formative assessment probes.* Arlington, VA: NSTA Press.

Section 1

Life and Its Diversity; Structure and Function; Life Processes and Needs of Living Things

Concept Matrix: Life and Its Diversity; Structure and Function; Life Processes and Needs of Living Things Probes #1–#13

PROBES	1. Cucumber Seeds	2. The Virus Debate	3. No Animals Allowed	4. Is It an Amphibian?	5. Pond Water	6. Atoms and Cells	7. Which One Will Dry Out Last?	8. Chlorophyll	9. Apple Tree	10. Light and Dark	11. Food for Corn	12. Pumpkin Seeds	13. Rocky Soil
GRADE-LEVEL USE →	K–8	6–12	K–8	K–8	3–12	6–12	9–12	6–12	3–8	6–12	6–12	K–8	K–8
RELATED CONCEPTS ↓													
amphibian				X									
animals			X	X	X								
atom						X							
bacteria							X						
behavior of organisms												X	X
cells	X	X			X	X	X	X					
cell shape							X						
cell structure					X								
characteristics of life	X	X											
chlorophyll								X	X				
chloroplast								X	X				
classification			X	X									
dormancy	X												
energy								X	X	X			
food									X		X		
germination	X											X	
leaf								X	X				
living	X	X				X							
organelles					X								
organs					X								
photosynthesis								X	X	X	X		
plants	X							X	X	X	X	X	X
reptile				X									
respiration	X									X		X	
seeds	X												
single-celled organism					X		X						
surface area							X						
tropism												X	X
virus		X											
volume							X						

Cucumber Seeds

Two friends bought packets of cucumber seeds. They argued about whether or not the cucumber seeds inside the sealed packets were living. Here is what they said:

Katie: "I think they are alive when they are in the sealed seed packet."

Vaughan: "I don't think they are alive until they are planted in the soil."

Which person do you agree with the most? ____________________

Explain your thinking.

Cucumber Seeds

Teacher Notes

Purpose

The purpose of this assessment probe is to elicit students' ideas about living things. The probe is designed to determine whether students recognize that seeds are living when they are in a dormant state.

Related Concepts

Living, seeds, plants, cells, germination, characteristics of life, respiration, dormancy

Explanation

The best answer is Katie's: "I think they are alive when they are in the sealed seed packet." Seeds are made of cells. When planted, the seeds in the packet germinate, give rise to seedlings, and grow into cucumber plants. For a living organism to grow, new cells must arise from other living cells. New cells do not arise from dead cells; therefore, the seed must be alive to germinate and grow into a seedling. If the seeds were not alive, they could not germinate, grow into seedlings, and eventually become mature plants able to produce their own seeds, keeping the life cycle going.

The seed is the beginning of a plant's life cycle. As a living thing, the seed requires energy to support metabolic activity, which it gets from the process of cellular respiration (taking in oxygen and breaking down molecules of food within the cell to release energy). However, in their dormant state, seeds require very little energy and thus cellular respiration in dormant seeds is negligible. When the seed begins to germinate, it requires much more energy to grow and cellular respiration speeds up significantly. Dormancy is a plant adaptation that blocks germination so that germination can occur when conditions are just right for the growth of a plant. The dormant stage is a living stage in which the metabolic activity of the seed is suspended. Seeds can survive for a long time in dormancy.

Curricular and Instructional Considerations

Elementary Students

In the elementary grades students learn about living things and life cycles. They develop an understanding that living things have basic needs such as food, water, and air. They observe functions performed by familiar living things such as eating, drinking, breathing, moving, and growing; however, a seed might not appear to be alive to them based on these observable characteristics. They have experiences growing seeds and observing plants go through their life cycles. It is important for students at this age to know that the seed is part of the life cycle of a plant and thus is living. Later in middle school, they will develop a deeper understanding of the characteristics of life that explain why seeds are considered living.

Middle School Students

As students investigate a variety of life forms and learn about life processes carried out by cells, they change or refine their earlier ideas about what is living and nonliving. Middle school students develop a deeper understanding of life processes such as cellular respiration, moving from a concept of breathing to understanding that respiration is a vital process organisms use to obtain the energy they need to live and grow. They learn that plants, animals, fungi, and most other organisms all depend on respiration to stay alive. They also learn that cells give rise to other living cells and that the process of cell division results in the growth of an organism such as a plant. They know that dead cells do not divide, grow, and contribute to the development of an organism. Middle school students can design simple investigations to obtain evidence that seeds are alive.

High School Students

In high school, students develop a more complete understanding of the metabolic processes that support life at both the cellular and molecular levels. High school students develop a more sophisticated understanding of aerobic and anaerobic respiration as a process ubiquitous to all life. They recognize death as the cessation of life processes and should be able to use logic and reason that if the seeds in the packet were dead, they could not grow into plants.

Administering the Probe

This probe is best used with K–8 students. Show students a packet of unopened seeds (or a picture of a seed packet) or a bag of dried beans from the grocery store to prompt their thinking about the seeds inside the packet or bag. Use the students' ideas and explanations to engage in a class discussion about what criteria determine whether something is living or nonliving. Listen carefully for students who think the seeds in the packet are dead but come alive when planted. This clearly indicates a need to help students understand death as the cessation of a life cycle.

Related Ideas in *National Science Education Standards* (NRC 1996)

K–4 Characteristics of Organisms

★ Organisms have basic needs. For example, animals need air, water, and food; plants require air, water, nutrients, and light.

K–4 Life Cycles of Organisms

★ Plants and animals have life cycles that include being born, developing into adults, reproducing, and eventually dying. The details of this life cycle are different for different organisms.

★ Indicates a strong match between the ideas elicited by the probe and a national standard's learning goal.

5–8 Structure and Function in Living Systems

- ★ All organisms are composed of cells—the fundamental unit of life.
- Cells carry out the many functions needed to sustain life. They grow and divide, therefore producing more cells. This requires that they take in nutrients, which they use to provide energy for the work that cells do and to make the materials that a cell or an organism needs.

9–12 Matter, Energy, and Organization in Living Systems

- ★ Living systems require a continuous input of energy to maintain their chemical and physical organizations. With death, and the cessation of energy input, living systems rapidly disintegrate.

Related Ideas in *Benchmarks for Science Literacy* (AAAS 2009)

K–2 The Cell

- ★ Most living things need water, food, and air.

3–5 The Cell

- Microscopes make it possible to see that living things are made mostly of cells.

6–8 The Cell

- Within cells, many of the basic functions of organisms—such as extracting energy from food and getting rid of waste—are carried out.
- The way in which cells function is similar in all living organisms.

9–12 The Cell

- Within the cells are specialized parts for the transport of materials, energy capture and release, protein building, waste disposal, passing information, and even movement.

Related Research

- Children have various ideas about what constitutes "living." Some may believe objects that are "active" are alive; for example, fire, clouds, or the Sun. As children mature, they include eating, breathing, and reproducing as essential characteristics of living things. People of all ages consider movement, and, in particular, movement in response to a stimulus, to be a defining characteristic of life. When doing so, these individuals tend to omit plants from the living category. Few young children give "growth" as a criterion for life, the exception being when plants are identified as living—then "growth" is commonly given as the reason (Driver et al. 1994).
- In a study of 424 Israeli students ages 8–14, there was no significant difference in age when students were asked to classify 16 pictures as living or nonliving things. Seeds were one of the most problematic items, although more students (60%) classified seeds as living than they did eggs. Some children believed eggs and seeds were not alive even when they held the idea that living things develop only from other living things (Tamir, Gal-Chappin, and Nussnovitz 1981).
- Some students may believe a seed only comes alive once it has been planted and begins to grow because they hold a conception that there are temporary breaks in an organism's life cycle that put it in a nonliving state. A related idea is that some students think caterpillars are alive but that when they go into the chrysalis state, they are nonliving and they become living again when they emerge as a butterflies (Allen 2010).

★ Indicates a strong match between the ideas elicited by the probe and a national standard's learning goal.

- A study by Stavy and Wax (1989) revealed that children seem to have different views for "animal life" and "plant life." In general, animals were more often recognized as being alive than plants.
- Elementary and middle school students use observable processes such as movement, breathing, reproducing, and dying when deciding if things are alive or not. High school and college students use these same readily observable characteristics to determine if something is alive. They rarely mention ideas such as "being made up of cells" or biochemical aspects such as "containing DNA." Many educators agree that the emphasis in science education on the learning of facts has contributed little toward understanding. Students may be able to quote the seven characteristics of life but may not be able to apply them when determining if something is living (Brumby 1982).
- Carey (1985) suggested that progression in the concept of *living* is linked to growth in children's ideas about biological processes. Young children have little knowledge of biology. In addition, it isn't until around the age of 9 or 10 years that children begin to understand death as the cessation of life processes.

Suggestions for Instruction and Assessment

- Combine this probe with "Is It Living?" (for elementary), "Is It Made of Cells?" (for middle and high school), and "Seedlings in a Jar" (for middle and high school) in *Uncovering Student Ideas in Science, Vol. 1: 25 Formative Assessment Probes* (Keeley, Eberle, and Farrin 2005).
- If students choose Vaughan's comment, challenge them to think of ways they can germinate seeds in the packet to test Vaughan's idea.
- If seed packets are unavailable, use grocery store beans. Most dried beans can be soaked and will germinate over two to three days. This might be more appropriate for children in urban settings who have had limited exposure to gardens and planting seeds.
- Place emphasis on the "living" aspect of the life cycle—every stage that an organism goes through is a living stage, all except for the final stage, death, and there is no life again after an organism dies. Because students often think organisms in the dormant state (such as seeds or the pupal stage of an insect) are dead, challenge them by asking how that could be if life cannot come from death.
- Distinguish needs of living things from the processes they use to sustain life. For example, a seed may not need water for many years while it is in a dormant state, as processes like cellular respiration are slowed down considerably. When environmental conditions are right and the seed takes in water, its respiration rate increases considerably as it uses energy to carry out its metabolic activities. In both cases the seed is alive, but the life processes are more obvious in the second situation.
- Many elementary teachers and students use the mnemonic MRS GREN to help students identify the seven observable processes that can be used to characterize life: M = movement (be aware that some organisms do not move during some stages in their life cycles), R = respiration, S = stimuli (react to), G = growth, R = reproduction, E = elimination of wastes, and N = nutrition (acquiring or making food). However, be aware that application of MRS GREN does little to convince students that seeds are alive, because they cannot observe most of these processes in a seed, except for growth, and that memorization of a list does not ensure conceptual understanding.

- For middle and high school students, consider using Miller and Levine's eight characteristics of living things: made up of cells, reproduction, made up of universal genetic code, growth and development, obtain and use materials for energy, response to environment, maintain stable internal environment, and change over time as a group (2006, p. 16).
- Use the metaphor of a hibernating animal, in which the animal's breathing rate and metabolism slow down significantly, to explain seed dormancy. The animal is still living, even though it appears to be barely breathing and does not eat during hibernation.
- High school students can measure the rate of respiration by comparing germinating and nongerminating seeds using respirometers or by building test tube respiration chambers that measure oxygen intake or release of carbon dioxide.

Related NSTA Science Store Publications, NSTA Journal Articles, NSTA SciGuides, NSTA SciPacks, and NSTA Science Objects

Ansberry, K., and E. Morgan. 2009. Teaching through trade books: Secrets of seeds. *Science and Children* 46 (6): 16–18.

Cavallo, A. 2005. Cycling through plants. *Science and Children* 42 (7): 22–27.

Konicek-Moran, R. 2008. Seed bargains. In *Everyday science mysteries: Stories for inquiry-based science teaching;* 107–144. Arlington, VA: NSTA Press.

Konicek-Moran, R. Forthcoming. Morning in the greenhouse. In *Yet more everyday science mysteries: Stories for inquiry-based science teaching.* Arlington, VA: NSTA Press.

Related Curriculum Topic Study Guides (in Keeley 2005)
"Cells"
"Characteristics of Living Things"
"Life Processes and Needs of Organisms"

References

American Association for the Advancement of Science (AAAS). 2008. Benchmarks for science literacy online. *www.project2061.org/publications/bsl/online*

Brumby, M. 1982. Students' perceptions of the concept of life. *Science Education* 66 (4): 613–622.

Carey, S. 1985. *Conceptual change in childhood.* Cambridge, MA: MIT Press.

Driver, R., A. Squires, P. Rushworth, and V. Wood-Robinson. 1994. *Making sense of secondary science: Research into children's ideas.* London: RoutledgeFalmer.

Keeley, P. 2005. *Science curriculum topic study: Bridging the gap between standards and practice.* Thousand Oaks, CA: Corwin Press and Arlington, VA: NSTA Press.

Keeley, P., F. Eberle, and L. Farrin. 2005. *Uncovering student ideas in science, vol. 1: 25 formative assessment probes.* Arlington, VA: NSTA Press.

Miller, K., and J. Levine. 2006. *Biology.* Upper Saddle River, NJ: Prentice-Hall.

National Research Council (NRC). 1996. *National science education standards.* Washington, DC: National Academies Press.

Stavy, R., and N. Wax. 1989. Children's conception of plants as living things. *Human Development* 32: 88–89.

Tamir, P., R. Gal-Chappin, and R. Nussnovitz. 1981. How do intermediate and junior high students conceptualize living and nonliving? *Journal of Research in Science Teaching* 18 (3): 241–248.

The Virus Debate

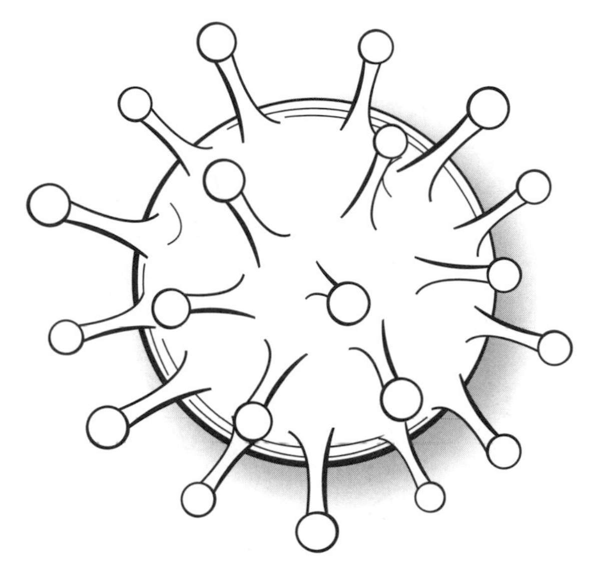

The students in Mrs. Oliver's biology class were having a debate about viruses. Half of the class thought viruses were considered living things. The other half of the class did not think viruses were considered living things. Which side are you on?

Circle the word you think best describes a virus: living nonliving

Explain your thinking. What rule or reasoning did you use to decide whether viruses are living or nonliving?

The Virus Debate

Teacher Notes

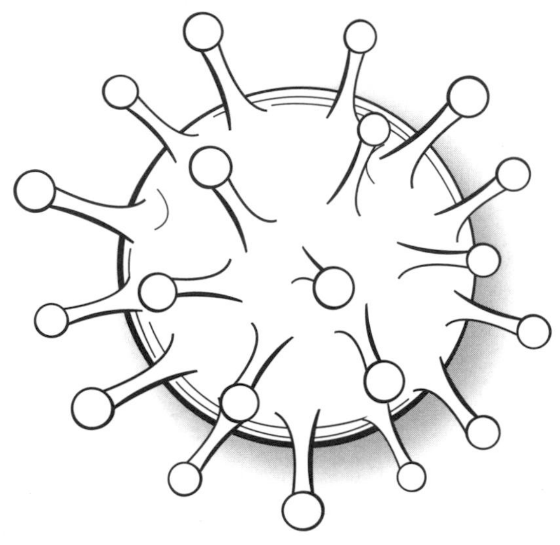

Purpose

The purpose of this assessment probe is to elicit students' ideas about characteristics of life. Viruses are used as a context to uncover students' ideas about what determines whether something is considered a living or a nonliving thing.

Related Concepts

Living, virus, cells, characteristics of life

Explanation

There is no absolutely right answer to this question, although the best answer leans toward nonliving. "Are viruses living?" is still an open-ended question in the scientific community and either answer is acceptable depending on students' reasoning. Viruses fall into a gray area between living and nonliving things. A virus is not a cell. It is just a piece of DNA or RNA (never both), surrounded by a protein coat called a capsid. Viruses are a thousand times smaller than even a bacterium and can exist in a wide variety of shapes and forms. They are often described as obligate intracellular parasites because they must be inside a living cell to reproduce and the relationship with its host is parasitic (the guest benefits at the expense of the host).

The reason the best answer leans toward nonliving is that when the characteristics of viruses are compared to a standard list of the characteristics of life, there are many more things they are *not*. Viruses are *not* made of cells, they do *not* obtain or use energy to run metabolic activities (they do *not* have a metabolism because they are just particles and not cells). They do *not* grow in size or develop over a lifetime from a juvenile virus to a mature virus. They do *not* have the ability to respond to a stimulus in their environment, and they do *not* maintain homeostasis as living cells do when they exchange gases, expel waste, or take in food and water.

So, why the debate? Well, viruses *do* have genes that can mutate and give the virus a new

characteristic that might allow it to have an advantage in its environment. Natural selection then "chooses" the viruses better able to infect new cells and thus "survive," which results in evolution of the virus population over time.

Also, viruses *do* "reproduce," but they are not capable of doing this independently, and they do not divide as cells do using mitosis or binary fission. Making more viruses is called replication rather than reproduction because they take over a living cell and use the cell's existing machinery to make copies of themselves by assembling more pieces of nucleic acids and protein coats. These new virus particles leave the host cell either one at a time or all at once to go infect a new host cell and make more copies. Whether a virus is living or nonliving, an organism or not an organism, leads to a lively debate during which students may conclude that viruses are at the borderline of life.

Curricular and Instructional Considerations

Elementary Students

In elementary grades, students learn about a variety of living things and some of the characteristics and needs that all living things have in common. They may be familiar with viruses as things that cause disease, but an understanding of the characteristics of a virus should wait until middle school when students have a deeper understanding of cells and life processes.

Middle School Students

In the middle grades, students develop a deeper understanding of the characteristics of life and know that the cell is the fundamental unit of structure and function that carries out the processes essential to all living things. Middle school students can understand the characteristics that define life and apply them to a variety of organisms. When viruses are introduced, they present a conundrum to students that challenges their conceptions of living versus nonliving. Middle school students develop a basic understanding of viruses and viral infection, but details about the types, structures, and replication of viruses can wait until high school.

High School Students

In high school, biology students learn about the history that led to the discovery of viruses. Their deeper knowledge of DNA, RNA, and proteins helps them understand the structural characteristics of viruses and the details of viral infection and replication, including several viruses of importance to human and plant health. Their understanding of the characteristics that define life at the cellular level deepens as they extend their knowledge to the level of cell organelles and biomolecules.

Administering the Probe

This probe is most appropriate for middle and high school students. Before giving the probe, make sure that students have some basic knowledge of what a virus is, its structure, and what it does when it enters a cell. After students have this basic information, use this probe to engage students in thinking about whether viruses are living. The probe leads to a lively debate among students, encouraging them to draw upon their prior knowledge of characteristics that define life.

Related Ideas in *National Science Education Standards* (NRC 1996)

K–4 Characteristics of Organisms

- Organisms have basic needs. For example, animals need air, water, and food; plants require air, water, nutrients, and light. Organisms can survive only in environments where their needs can be met.

5–8 Structure and Function in Living Systems

- ★ All organisms are composed of cells—the fundamental unit of life.
- ★ Cells carry out the many functions needed to sustain life. They grow and divide, therefore producing more cells. This requires that they take in nutrients, which they use to provide energy for the work that cells do and to make the materials that a cell or an organism needs.

5–8 Reproduction and Heredity

- ★ Reproduction is a characteristic of all living systems.

9–12 The Cell

- Cells have particular structures that underlie their functions. Every cell is surrounded by a membrane that separates it from the outside world. Inside the cell is a concentrated mixture of different molecules that form a variety of specialized structures that carry out such cell functions as energy production, transport of molecules, waste disposal, synthesis of new molecules, and the storage of genetic material.
- Cells store and use information to guide their functions. The genetic information stored in DNA is used to direct the synthesis of the thousands of proteins that each cell requires.
- Cell functions are regulated. Regulation occurs through both changes in the activity of the functions performed by proteins and through the selective expression of individual genes. This regulation allows cells to respond to their environment and to control and coordinate cell growth and division.

9–12 The Molecular Basis of Heredity

- In all organisms, the instructions for specifying the characteristics of the organism are carried in DNA, a large polymer formed from subunits of four kinds (A, G, C, and T). The chemical and structural properties of DNA explain how the genetic information that underlies heredity is both encoded in genes and replicated.

Related Ideas in *Benchmarks for Science Literacy* (AAAS 2009)

K–2 The Cell

- Most living things need water, food, and air.

3–5 The Cell

- Microscopes make it possible to see that living things are made mostly of cells.

6–8 The Cell

- ★ All living things are composed of cells, from just one to many millions, whose details usually are visible only through a microscope.
- ★ Within cells, many of the basic functions of organisms—such as extracting energy from food and getting rid of waste—are carried out.
- Cells repeatedly divide to make more cells for growth and repair.

9–12 The Cell

- Within the cells are specialized parts for the transport of materials, energy capture and release, protein building, waste disposal, passing information, and even movement.

Related Research

- Children have various ideas about what constitutes "living." Some may believe objects that are "active" are alive; for example, fire, clouds, or the Sun. As children mature, they include eating, breathing, and reproducing as essential characteris-

★ Indicates a strong match between the ideas elicited by the probe and a national standard's learning goal.

tics of living things. People of all ages use movement and, in particular, movement in response to a stimulus, as a defining characteristic of life (Driver et al. 1994).

- Elementary and middle school students use observable processes such as movement, breathing, reproducing, and dying when deciding if things are alive or not. High school and college students use these same readily observable characteristics to determine if something is alive. They rarely mention ideas such as "being made up of cells" or biochemical aspects such as "containing DNA." Many educators suggested that the learning of facts has contributed little toward understanding. Students may be able to quote the seven characteristics of life but may not be able to apply them when determining if something is living (Brumby 1982).
- Carey (1985) suggested that progression in the concept of *living* is linked to growth in children's ideas about biological processes. Young children have little knowledge of formal concepts in biology, including criteria that define characteristics of life.

Suggestions for Instruction and Assessment

- Combine this probe with "Is It Living?" and "Is It Made of Cells?" in *Uncovering Student Ideas in Science, Vol. 1: 25 Formative Assessment Probes* (Keeley, Eberle, and Farrin 2005). Consider adding *virus* to the list of distracters in each of these probes.
- This probe can also be used with the "agreement circle" formative assessment classroom technique (FACT) (Keeley 2008). Students are asked several questions about viruses while standing in a large circle. Those who agree with the statement step to the inside of the circle. Those who disagree stay on the outside of the circle. Then students from inside and outside the circle are matched up in groups to argue their ideas. After using several questions about viruses, use this probe as the culminating question for debate.
- Use graphic organizers such as Venn diagrams or compare-and-contrast charts to compare characteristics of viruses to characteristics of living cells.
- Miller and Levine (2006, p. 480) use an interesting analogy to describe lytic infection by viruses. Challenge students to compare viral infection to an outlaw in the American Wild West—the virus being the outlaw and the cell being the town. First, the outlaw rides into town and then eliminates the town's existing authority (the sheriff), which is analogous to the host cell's DNA. Then the outlaw demands to be outfitted with new weapons, horses, and equipment by terrorizing the local people, which is analogous to using the host cell to make viral proteins and viral DNA. Finally, the outlaw forms a gang that leaves the town and goes on to attack new towns, which is analogous to the host cell bursting open and releasing hundreds of virus particles that go on to infect new cells.
- Ask students if viruses are parasites. Compare viruses to parasitic organisms. How are they alike and how are they different? How can a virus be considered a parasite, yet not be considered an organism?
- Compare and contrast how viruses are classified with how living organisms are classified.

Related NSTA Science Store Publications, NSTA Journal Articles, NSTA SciGuides, NSTA SciPacks, and NSTA Science Objects

Dubois, S., K. Moulton, and J. Jamison. 2009. Collaboration at the nanoscale: Exploring viral genetics with electron microscopy. *The Science Teacher* 76 (4): 32–39.

Jones, M., M. Falvo, A. Taylor, and B. Broadwell. 2007. *Nanoscale science: Activities for grades 6–12.* Arlington, VA: NSTA Press.

Diamond, J., with C. Zimmer, E. Evans, L. Allison, and S. Disbrow. 2006. *Virus and the whale: Exploring evolution in creatures large and small.* Arlington, VA: NSTA Press.

Related Curriculum Topic Study Guide (in Keeley 2005)
"Characteristics of Living Things"

References

American Association for the Advancement of Science (AAAS). 2009. Benchmarks for science literacy online. *www.project2061.org/publications/bsl/online*

Brumby, M. 1982. Students' perceptions of the concept of life. *Science Education* 66 (4): 613–622.

Carey, S. 1985. *Conceptual change in childhood.* Cambridge, MA: MIT Press.

Driver, R., A. Squires, P. Rushworth, and V. Wood-Robinson. 1994. *Making sense of secondary science: Research into children's ideas.* London: RoutledgeFalmer.

Keeley, P. 2005. *Science curriculum topic study: Bridging the gap between standards and practice.* Thousand Oaks, CA: Corwin Press and Arlington, VA: NSTA Press.

Keeley, P. 2008. *Science formative assessment: 75 practical strategies for linking assessment, instruction, and learning.* Thousand Oaks, CA: Corwin Press and Arlington, VA: NSTA Press.

Keeley, P., F. Eberle, and L. Farrin. 2005. *Uncovering student ideas in science, vol. 1: 25 formative assessment probes.* Arlington, VA: NSTA Press.

Miller, K., and J. Levine. 2006. *Biology.* Upper Saddle River, NJ: Prentice-Hall.

National Research Council (NRC). 1996. *National science education standards.* Washington, DC: National Academies Press.

No Animals Allowed

Susan and Mario saw a sign on a store that read "No animals allowed." They wondered if people were considered animals. This is what they said:

Susan: "I think the sign should be changed. People are animals."

Mario: "I think the sign is OK. People are not animals. They are humans."

Whom do you agree with the most? ____________ Explain your thinking about what an animal is.

____

____

____

____

____

____

____

____

____

No Animals Allowed

Teacher Notes

Purpose

The purpose of this assessment probe is to elicit students' ideas about biological classification. The probe is designed to determine whether students recognize humans as animals.

Related Concepts

Classification, animals

Explanation

The best answer is Susan's: "I think the sign should be changed. People are animals." Animals share several characteristics that set them apart from other organisms. All animals are multicellular eukaryotes. Their cells lack cell walls and are made up of membrane-bound organelles. Animals are heterotrophs, meaning they must obtain their own food by eating other organisms. Animals are generally motile and move voluntarily. Embryological development is also used to characterize animals. During early development, animal embryos pass through a blastula stage (a hollow ball of cells that develops after the fertilized egg divides). Because humans fit all of these characteristics of an animal, biologically they are considered animals. Overall, animals are diverse organisms with examples ranging from, but not limited to, sponges, jellyfish, worms, insects, spiders, fish, lizards, frogs, and humans. Humans belong in the same taxonomic class as dogs, cats, and other mammals.

Curricular and Instructional Considerations

Elementary Students

By the time children enter school, they have encountered a variety of animals. They have begun to develop their own operational definitions of an animal. Their operational definitions are often reinforced by school experiences that use farm, zoo, backyard, school yard, and family pets as examples of animals. By the intermediate grades, students should have opportunities to observe and describe a

variety of animals, including invertebrates, and understand some of the basic characteristics and needs they all have in common. They may begin to invent their own schemes for classifying animals into further categories, based on similar characteristics. Teachers who start with the students' own approaches, rather than with formal classification schemes, help elementary students realize that taxonomic classification is used to understand relatedness among organisms and that how animals are classified depends on the purpose of classification.

Middle School Students

In middle school, students develop formal distinctions between animals and other eukaryotes (organisms with membrane-bound organelles in their cells) based on how they acquire their food and on their internal and external anatomical characteristics. Middle school students recognize animals as consumers before they are introduced to the word *heterotroph*. Although biological diversity is addressed at this level, some students may think humans are in their own separate category, even though they recognize that humans belong to the phylum of vertebrates and the class Mammalia.

High School Students

Although classification has been de-emphasized in the science standards at this level, students exhibit a general understanding of taxonomic classification and use hierarchical groupings to understand that seemingly different organisms belong to the animal kingdom. Although Linnaeus's binomial scientific names are still used, the old five-kingdom approach (animals, plants, fungi, protists, monera) is being replaced with the modern approach using domains. The three domains more accurately reflect the evolutionary history of life—Eubacteria, Archaebacteria, and Eukarya (which animals belong to).

Administering the Probe

This probe is designed for grades K–8 but can be used at all grade levels to determine whether students recognize humans as animals. It is particularly useful in a discussion format where students argue and defend their ideas about whether or not humans are considered animals.

Related Ideas in *National Science Education Standards* (NRC 1996)

K–4 Characteristics of Organisms

- Organisms have basic needs. For example, animals need air, water, and food; plants require air, water, nutrients, and light.
- Each plant or animal has different structures that serve different functions in growth, survival, and reproduction.

5–8 Diversity and Adaptations of Organisms

★ Millions of species of animals, plants, and microorganisms are alive today. Although different species might look dissimilar, the unity among organisms becomes apparent from an analysis of internal structures, the similarity of their chemical processes, and the evidence of common ancestry.

9–12 Biological Evolution

★ Biological classifications are based on how organisms are related. Organisms are classified into a hierarchy of groups and subgroups based on similarities that reflect their evolutionary relationships.

Related Ideas in *Benchmarks for Science Literacy* (AAAS 2009)

K–2 Diversity of Life

★ Some animals and plants are alike in the way they look and in the things they do,

★ Indicates a strong match between the ideas elicited by the probe and a national standard's learning goal.

and others are very different from one another.

3–5 Diversity of Life

- A great variety of kinds of living things can be sorted into groups in many ways using various features to decide which things belong to which group.
- There are millions of different kinds of individual organisms that inhabit the Earth at any one time—some very similar to each other, some very different.

6–8 Diversity of Life

★ One of the most general distinctions among organisms is between plants, which use sunlight to make their own food, and animals, which consume energy-rich foods. Some kinds of organisms, many of them microscopic, cannot be neatly classified as either plants or animals.

- Similarities among organisms are found in internal anatomical features, which can be used to infer the degree of relatedness among organisms.
- In classifying organisms, scientists consider details of both internal and external structures.

9–12 Diversity of Life

- Similar patterns of development and internal anatomy suggest relatedness among organisms.

★ A classification system is a framework created by scientists for describing the vast diversity of organisms, indicating the degree of relatedness between organisms, and framing research questions.

Related Research

- People of all ages have a much narrower definition of an animal than biologists. Typically, younger students think of "animals" as large, terrestrial mammals. Animal characteristics commonly include the following: having four legs, being large in size, having fur, making noise, and living on land. A study of 15-year-olds found that only 10% identified animals as a biologist would from an assortment of organisms (Bell 1981).
- Humans are often not thought of as animals; rather they are contrasted with animals and thought of as an alternative to *animal* rather than a subset (Driver et al. 1994).
- Elementary and middle school students hold a much more restricted meaning for "animal" than the definition a biologist would use (Mintzes et al. 1991).
- Studies show that preservice elementary teachers, as well as experienced elementary teachers, also hold restricted meanings for the concept of *animal.* This may affect students' opportunities to learn the scientific meaning of *animal* (Driver et al. 1994).
- Some research indicates that in second grade there is a shift in children's understanding of organisms from representations based on perceptual and behavioral features to representations in which central principles of biological theory are most important (AAAS 2007, p. 30).
- By middle school, students begin to group organisms hierarchically when prompted to do so. In high school, students are more apt to use hierarchical taxonomies without prompting (Leach et al. 1992).

Suggestions for Instruction and Assessment

- Combine this probe with "Is It an Animal?" in *Uncovering Student Ideas in Science, Vol. 1: 25 Formative Assessment Probes* (Keeley, Eberle, and Farrin 2005).
- Ask students to write down what they think something has to be like before it can be considered an animal. Also, ask students to draw an animal. Note how many

★ Indicates a strong match between the ideas elicited by the probe and a national standard's learning goal.

of their drawings match what the research predicts students think an animal is.

- Once students have developed an operational definition of an animal and compared it to a scientific definition appropriate for their grade level, they should be encouraged to further group animals into categories of "animal." By developing their own classification schemes to further group different types of animals, they will see that animals are a diverse group of organisms, and although they may have special characteristics within a group, they all share certain common characteristics. For older students, their own schemes can be compared with schemes used by scientists.
- When students group organisms as animals, it is important that instruction is geared toward getting students to carefully examine the characteristics used for grouping to see if they are truly exclusive.
- When deciding how to group humans, if students do not believe they are grouped as animals, ask students if they would be placed in any of the remaining kingdoms. When they see that they cannot be placed with plants, animals, fungi, protists (or protoctista), or simple prokaryotes (bacteria and other simple one-celled organisms), they may accept "animal" by default. Students can then examine the animal characteristics to further understand why humans are grouped as animals.
- Be aware of how children's picture books, references to animals in the media, toys such as stuffed animals, and other everyday uses of the word *animal* limit students' understanding of the precise biological meaning of the word *animal.*
- Elementary instruction related to animals is often focused on a specific organism or organisms. This may help students develop an understanding of the special characteristics of that organism(s) that define it as an animal, but it may fail to help students apply the generalized notion of *animal* to other organisms. Be sure to explicitly develop generalizations when focusing on one type of organism in a lesson.
- Encourage students to examine different features of animals. Identify attributes in common, even though the animals appear very different from one another.
- Help middle school students develop the skills of hierarchical grouping. For example, a human is a mammal, a vertebrate, and an animal. Consider introducing advanced students to the concept of flexible clades rather than using strict categories as a way to classify organisms hierarchically.
- Help students understand that the way we use words in everyday life is often different from the way we use words in science. *Animal* has a much more precise meaning in science than in our everyday language. Point out other examples of words that often differ between the way they are used every day and the way they are used in science, such as *theory*, *adapt*, and *food.*
- Use the interview protocol developed by Charles Barman to further probe for students' ideas about animals (Barman et al. 1999). This interview protocol is described in the NSTA *Science and Children* member journal included in the NSTA resource list on page 26.
- Be aware that accepting humans as animals may challenge students' cultural or religious beliefs that make them resistant to the idea that humans are animals. In this case, it is important for students to know that scientists have a specific definition for an "animal," and that humans fit this definition. Teachers can respect students' beliefs by balancing the scientific notion that humans are biologically classified as animals with the notion that humans are a unique and very special kind of animal.

- An excellent web resource is the Tree of Life Project at *http://tolweb.org/tree.*

Related NSTA Science Store Publications, NSTA Journal Articles, NSTA SciGuides, NSTA SciPacks, and NSTA Science Objects

American Association for the Advancement of Science (AAAS). 2007. *Atlas of science literacy.* Vol. 2. (See "Diversity of Life" map, pp. 30–31.) Washington, DC: AAAS.

Barman, C., N. Barman, K. Berglund, and M. Goldston. 1999. Assessing students' ideas about animals. *Science and Children* 37 (1): 44–49.

Stovall, G., and C. Nesbit. 2003. Let's try action research. *Science and Children* 40 (5): 44–48.

Related Curriculum Topic Study Guides (in Keeley 2005)
"Animal Life"
"Biological Classification"

References

American Association for the Advancement of Science (AAAS). 2007. *Atlas of science literacy.* Vol. 2. Washington, DC: AAAS.

American Association for the Advancement of Science (AAAS). 2009. Benchmarks for science literacy online. *www.project2061.org/publications/bsl/online*

Barman, C., N. Barman, K. Berglund, and M. Goldston. 1999. Assessing students' ideas about animals. *Science and Children* 37 (1): 44–49.

Bell, B. 1981. When is an animal not an animal? *Journal of Biological Education* 15 (3): 213–218.

Driver, R., A. Squires, P. Rushworth, and V. Wood-Robinson. 1994. *Making sense of secondary science: Research into children's ideas.* London: RoutledgeFalmer.

Keeley, P. 2005. *Science curriculum topic study: Bridging the gap between standards and practice.* Thousand Oaks, CA: Corwin Press and Arlington, VA: NSTA Press.

Keeley, P., F. Eberle, and L. Farrin. *2005. Uncovering student ideas in science, vol. 1: 25 formative assessment probes.* Arlington, VA: NSTA Press.

Leach, J., R. Driver, P. Scott, and C. Wood-Robinson. 1992. *Progression in understanding of ecological concepts by pupils aged 5 to 16.* Leeds, UK: University of Leeds, Centre for Studies in Science and Mathematics Education.

Mintzes, J., J. Trowbridge, M. Arnaudin, and J. Wandersee. 1991. Children's biology: studies on conceptual development in the life sciences. In *The psychology of learning science,* ed. S. Glynn, R. Yeany, and B. Britton, 179–202. Hillsdale, NJ: Lawrence Erlbaum Associates.

National Research Council (NRC). 1996. *National science education standards.* Washington, DC: National Academies Press.

Is It an Amphibian?

Belinda's sister came home excited about her science class. They were studying amphibians. Her sister asked Belinda to help her make a list of amphibians.

Put an X next to the animals they should put on their list.

_____ tree frog	_____ dragonfly	_____ pond turtle
_____ water snake	_____ penguin	_____ beaver
_____ shark	_____ alligator	_____ bullfrog
_____ duck	_____ salamander	_____ whale
_____ catfish	_____ toad	_____ sea turtle
_____ mosquito	_____ crab	_____ seal
_____ snail	_____ rattlesnake	_____ eel

Explain your thinking. What rule or reasoning did you use to decide if something is an amphibian?

____

____

____

____

____

____

____

Is It an Amphibian?

Teacher Notes

Purpose

The purpose of this assessment probe is to elicit students' ideas about a class of vertebrate animals—amphibians. The probe uses a "justified list" format (Keeley 2008, p. 123) to determine how students decide whether an animal is considered an amphibian. It reveals whether students use the general criterion that most amphibians are adapted to spending parts of their life cycles in water and parts on land.

Related Concepts

Classification, animals, amphibian, reptile

Explanation

There are four amphibians on the list: tree frog, bullfrog, salamander, and toad. Amphibians are sometimes erroneously called "cold-blooded" but are labeled by scientists as ectothermic or poikilothermic (a more accurate word) because their internal temperatures are dependent on the temperature of their surroundings. They are a smooth-skinned class of vertebrate animals adapted to spend part of their life cycles on land and part in water. They characteristically hatch as aquatic larvae with gills and spend part of their life cycles in water (although some, such as the axolotl, spend their entire lives in water). The larvae then transform into adults having air-breathing lungs, enabling them to live on land (e.g., tadpole to frog) until it's time to reproduce. Then most return to freshwater to lay their shell-less eggs. When on land, amphibians require moist habitats, because they rely heavily on their moist skin for gas exchange. Some species lack lungs and exchange gases exclusively through their skin and mouths.

Curricular and Instructional Considerations

Elementary Students

Young children are fascinated by familiar amphibians such as toads and frogs. Many younger students have watched tadpoles transform into frogs or toads, and many have had

experiences with newts and other familiar salamanders. When they learn about life cycles, the life cycle of a frog is used as a common example that shows how the young animal completes part of its life cycle in water as a gill-breathing animal without legs and gradually transforms into an adult with lungs and legs that is able to live on land. By the intermediate grades, students learn that these animals are classified as amphibians.

Middle School Students

In middle school, students develop formal distinctions between classes of vertebrate animals as well as invertebrates that live on land, water, or both. Middle school students are more apt to use juvenile and adult anatomical characteristics to classify amphibians rather than the general definition that amphibians live part of their lives in water and part on land, which can apply to insects and other organisms.

High School Students

Although classification has been de-emphasized in the standards at this level, students exhibit a general understanding of taxonomic classification and use hierarchical groupings to understand that seemingly different vertebrate animals, such as amphibians, belong to the same class based on common characteristics, and that within those classes there are further subgroupings such as orders, families, genera, and species.

Administering the Probe

This probe is best used at grades K–8 to determine whether students recognize amphibians as a distinct group (class) of vertebrate animals. It can also be used with high school students to reveal whether they still have misconceptions about amphibians. Make sure students are familiar with the organisms on the list on page 27. Remove any unfamiliar organisms from the probe. For English language learners or developing readers, you may include pictures with names. This probe works well as a card sort, one of the FACTs (formative assessment classroom techniques) used with probes in the form of a list (Keeley 2007). Write the name of each organism on a separate card. Make four or five sets of cards. Have students work in pairs or small groups to sort the cards into two piles: those they think are amphibians and those they think are something else. Remind students to explain their reasoning as they sort each card. As you circulate among the class, notice which cards are placed in the amphibian category, listen to students explain their reasoning, and probe further to find out why they placed some of the organisms that are not amphibians in the amphibian category. This probe can be used as a whole-class discussion prior to instruction to elicit the class' conception of what an amphibian is. After students have had opportunities to develop the formal conception of what an amphibian is, the list can be revisited to see if students have changed any of their ideas.

Related Ideas in *National Science Education Standards* (NRC 1996)

K–4 Characteristics of Organisms

- ★ Each plant or animal has different structures that serve different functions in growth, survival, and reproduction.

K–4 Life Cycles of Organisms

- Plants and animals have life cycles that include being born, developing into adults, reproducing, and eventually dying. The details of this life cycle are different for different organisms.

5–8 Diversity and Adaptations of Organisms

- Millions of species of animals, plants, and microorganisms are alive today. Although

★ Indicates a strong match between the ideas elicited by the probe and a national standard's learning goal.

different species might look dissimilar, the unity among organisms becomes apparent from an analysis of internal structures, the similarity of their chemical processes, and the evidence of common ancestry.

9–12 Biological Evolution

- ★ Biological classifications are based on how organisms are related. Organisms are classified into a hierarchy of groups and subgroups based on similarities that reflect their evolutionary relationships.

Related Ideas in *Benchmarks for Science Literacy* (AAAS 2009)

K–2 Diversity of Life

- ★ Some animals and plants are alike in the way they look and in the things they do, and others are very different from one another.

3–5 Diversity of Life

- ★ A great variety of kinds of living things can be sorted into groups in many ways using various features to decide which things belong to which group.
- There are millions of different kinds of individual organisms that inhabit the Earth at any one time—some very similar to each other, some very different.

6–8 Diversity of Life

- ★ In classifying organisms, scientists consider details of both internal and external structures.

9–12 Diversity of Life

- Similar patterns of development and internal anatomy suggest relatedness among organisms.
- ★ A classification system is a framework created by scientists for describing the vast diversity of organisms, indicating the degree of relatedness between organisms, and framing research questions.

Related Research

- A common misconception among children in England and Wales is that lizards are amphibians. This may be attributed to the similar appearance of lizards and salamanders in pictures they have seen (Allen 2010).
- When elementary, middle, and high school students were asked to sort amphibians and reptiles, most students were able to classify snakes as reptiles. However, 70% of students (across all grades sampled), classified sea turtles as amphibians. Interestingly, the alternative views of sea turtles as amphibians remained intact throughout the school years (Yen, Yao, and Chin 2004).
- A study of third-grade teachers' knowledge of animal classification showed misconceptions and lack of understanding similar to elementary students. Few teachers in the study were able to explain that amphibians breathe through gills early in their lives, and then breathe through lungs once the metamorphosis has taken place. Forty percent (n = 41) believed a crab is an amphibian, while 24% thought a sea turtle was an amphibian. A wide range of misconceptions surfaced related to distinguishing between amphibians and reptiles. Some believed that reptiles have moist skin so that they do not dry out and die (which is actually a characteristic of amphibians, as they need moist skin for gas exchange) (Bucher, Burgoon, and Duran 2010).
- Some research indicates that in second grade there is a shift in children's understanding of organisms from representations based on perceptual and behavioral features to representations in which central principles of biological theory are most important (AAAS 2007, p. 30).

★ Indicates a strong match between the ideas elicited by the probe and a national standard's learning goal.

- By middle school, students begin to group organisms hierarchically when prompted to do so. In high school, students are more apt to use hierarchical taxonomies without prompting (Leach et al. 1992).

Suggestions for Instruction and Assessment

- Combine this probe with "Is It an Animal?" in *Uncovering Student Ideas in Science, Vol. 1: 25 Formative Assessment Probes* (Keeley, Eberle, and Farrin 2005).
- Ask students to write down what they think something has to be like before it can be considered an amphibian. Then ask students to draw an amphibian. Note whether any of their drawings show reptiles or other organisms.
- The term *cold-blooded* is seldom used in biology today. The more appropriate term is *ectothermic,* which means an organism relies on interactions with its environment to control its body temperature. With younger students, it may be better to develop the concept of *ectothermic* before introducing the word itself. For example, teach students that some animals, such as amphibians, cannot maintain a steady body temperature when the environment changes.
- Once students have developed an understanding that animals can be grouped as vertebrates or invertebrates, they can be encouraged to further group animals into classes of vertebrates using their own schemes. By developing their own classification schemes to further group different types of vertebrates, they will see that vertebrates are a diverse group of organisms. Once students have developed their schemes, they can be compared with the classes of vertebrates, including amphibians, used by scientists.
- Be aware that vaguely describing amphibians as animals that can live on both land and in water may lead to students selecting organisms such as alligators, beavers, penguins, turtles, crabs, and certain types of insects (e.g., mosquito larvae live in water and adult mosquitoes live on land), from the list on page 27 . When developing the concept of an amphibian, be sure that students become familiar with the criteria for identifying an organism as an amphibian: (1) it relies on interactions with its environment to control its body temperature (ectothermic); (2) it has moist skin; (3) it has a unique limb structure; (4) it is a vertebrate; and (5) it has, at different stages in its life cycle, either gills or lungs (*Note:* Some amphibians such as the axolotl remain aquatic and breathe through gills their entire lives).
- Everyday descriptions such as "amphibious vehicle" may perpetuate the idea that amphibians are organisms that can go in and out of water, such as turtles, alligators, and penguins.
- Students who live in northern climates may have never seen a live lizard but have seen salamanders, and students who live in desert areas may have seen lizards but not salamanders. Pictures of a salamander and a common lizard, such as an anole, look very similar. Have students compare the visual similarities and help them construct an understanding of why one is considered an amphibian and the other is a reptile. Older students can also compare internal anatomical differences, such as a three-chambered heart in amphibians and a four-chambered heart in reptiles, and the difference in limb structure.

Related NSTA Science Store Publications, NSTA Journal Articles, NSTA SciGuides, NSTA SciPacks, and NSTA Science Objects

Alexander, D. 2010. *Hop into action: The curriculum guide for grades K–4.* Arlington, VA: NSTA Press.

American Association for the Advancement of Science (AAAS). 2007. *Atlas of science literacy*. Vol. 2. (See "Diversity of Life" map, pp. 30–31.) Washington, DC: AAAS.

ReMine, S. 2005. Science sampler: Learning from amphibians. *Science Scope* 29 (3): 60–61.

Schneider, R., M. Krasny, and S. Moreale. 2001. *Hands-on herpetology: Exploring ecology and conservation.* Arlington, VA: NSTA Press.

Related Curriculum Topic Study Guides (in Keeley 2005)
"Animal Life"
"Biological Classification"
"Reproduction, Growth, and Development (Life Cycles)"

References

Allen, M. 2010. *Misconceptions in primary science.* Berkshire, England: Open University Press.

American Association for the Advancement of Science (AAAS). 2007. *Atlas of science literacy.* Vol. 2. Washington, DC: AAAS.

American Association for the Advancement of Science (AAAS). 2009. Benchmarks for science literacy online. *www.project2061.org/publications/bsl/online*

Bucher, A., J. Burgoon, and E. Duran. 2010. Exploring elementary and middle school inservice teachers' knowledge of animal classification: A comparison of student and teacher misconceptions. *National Social Science Journal* 35 (1): 31–37.

Keeley, P. 2005. *Science curriculum topic study: Bridging the gap between standards and practice.* Thousand Oaks, CA: Corwin Press and Arlington, VA: NSTA Press.

Keeley, P. 2008. *Science formative assessment: 75 practical strategies for linking assessment, instruction, and learning.* Thousand Oaks, CA: Corwin Press and Arlington, VA: NSTA Press.

Keeley, P., F. Eberle, and L. Farrin. 2005. *Uncovering student ideas in science, vol. 1: 25 formative assessment probes.* Arlington, VA: NSTA Press.

Leach, J., R. Driver, P. Scott, and C. Wood-Robinson. 1992. *Progression in understanding of ecological concepts by pupils aged 5 to 16.* Leeds, UK: University of Leeds, Centre for Studies in Science and Mathematics Education.

National Research Council (NRC). 1996. *National science education standards.* Washington, DC: National Academies Press.

Yen, C., T. Yao, and Y. Chiu. 2004. Alternative conceptions in animal classification focusing on amphibians and reptiles: A cross-age study. *International Journal of Science and Mathematics Education* 2 (2): 159–174.

Pond Water

Six students were looking at a drop of pond water through a microscope. They were amazed to see many different types of tiny single-celled organisms moving around in the drop of water. The students wondered what they would see if they had a more powerful microscope. They wondered how the insides of the single-celled organisms compared to the insides of animals. This is what they said:

Lanny: "I don't think they have any of the organs animals have."

Dorothy: "I think they have a few of the organs most animals have."

Seamus: "I think they have a few of the organs simple animals, like worms, have."

Nick: "I think the only organs they have that animals also have are the digestive organs."

Valynda: "I think the only organs they have that animals also have are the organs they use for breathing."

Brian: "I think they have all of the organs that most animals have, they are just a lot smaller."

Which person do you agree with the most? _______________ Explain why you agree.

Describe what you think you would see if you could look inside a single-celled organism with a powerful microscope.

____

____

____

Pond Water

Teacher Notes

Purpose

The purpose of this assessment probe is to determine how students distinguish the internal structure of single-celled organisms from multicelled organisms, such as animals. The probe is designed to reveal whether students recognize that single-celled organisms do not contain "tiny versions" of familiar organs found in animals.

Related Concepts

Cells, single-celled organism, animals, organs, organelles, cell structure

Explanation

The best answer is Lanny's: "I don't think they have any of the organs animals have." Organs found in animals differ structurally from organelles found in single-celled organisms. By definition, an organ is a group of tissues that work together to perform a similar function. The heart is an example of an organ. Tissues are groups of similar cells that perform a similar function, such as the muscle tissue found in the heart. Organs are only found in multicellular organisms because they are made up of many cells; therefore, single-celled organisms cannot contain organs. Their unit of structure is called the organelle, a specialized unit of structure and function found within eukaryotic cells. Translated, the word *organelle* means tiny organ. However, organelles are not tiny versions of organs found in animals. They are a subset of a cell, made up of biomolecules such as proteins, carbohydrates, lipids, and nucleic acids.

Although single-celled organisms and animals perform similar life functions such as acquiring food, releasing energy from food, responding to the environment, and getting rid of wastes, these functions are organized hierarchically in animals from the organ-system level down to the level of the cell. Even simple animals such as worms contain tissues arranged in organs. Single-celled organisms lack this hierarchal grouping of cells. Their life functions are

performed by structures within their cell and do not involve organization of cells from cell to tissue to organ to organ system.

Curricular and Instructional Considerations

Elementary Students

In the upper elementary grades, students are introduced to single-celled organisms and given opportunities to collect pond water and observe the various types of microorganisms in a drop of water. The focus at this level is on the diversity of different types of organisms and on understanding that they have similar needs and carry out the same life processes as multicellular plants and animals. Students also learn about organs and organ systems when they study the human body. Elementary students should know that single-celled organisms are made up of a single cell and, because of this, their "body structures" differ from organisms such as plants and animals that have organs that perform different functions for the organism. However, details of the internal structure of eukaryotic microorganisms can wait until middle and high school.

Middle School Students

Middle school students develop a hierarchical understanding of the organization of multicellular organisms, particularly as they study the human body. They learn that cells work together to form tissues and tissues group together to form organs. They build on their previous observations of different types of microscopic organisms they collect in their environments to understand how these organisms are classified biologically, and they are introduced to the structures of the organisms. They begin to recognize the similarities between the organelles of single-celled protists and organelles in the cells of animals. For example, they recognize that an amoeba and a skin cell both have a nucleus and are surrounded by a cell membrane. Middle school students should recognize that organelles are not the same as organs.

High School Students

In high school, students explore in detail the structure and functions of cell organelles, including the organelles found in single-celled eukaryotic organisms. They use images from electron micrographs to reveal details of cell organelles. High school students develop a hierarchical understanding of structure and function at the microscopic level going from the cell to cell organelles to the molecules that function within organelles.

Administering the Probe

This probe is most appropriate beginning at grade 3. Once students have had experiences learning about human or animal body organs, it can be used as an elicitation of their ideas before they are introduced to single-celled organisms. The probe can be used with middle and high school students to examine whether they use hierarchical reasoning to compare single-celled and multicellular organisms. Even though high school biology students have previously learned about cells, research shows that some students still cling to the misconception that there are "tiny organs," similar to human organs, within cells. Consider combining this probe with a graphic showing different types of organisms in a drop of pond water. However, be aware that when students look at local pond water samples they frequently see animals such as rotifers, arthropods, and small worms. These organisms do have organs. Juveniles of these multicellular organisms often are no bigger than the larger unicellular organisms found in the same sample.

Related Ideas in *National Science Education Standards* (NRC 1996)

K–4 The Characteristics of Organisms

- Organisms have basic needs.

5–8 Structure and Function in Living Systems

★ All organisms are composed of cells—the fundamental unit of life. Most organisms are single cells; other organisms, including humans, are multicellular.

★ Important levels of organization for structure and function include cells, organs, tissues, organ systems, whole organisms, and ecosystems.

- Groups of specialized cells cooperate to form a tissue, such as a muscle. Different tissues are in turn grouped together to form larger functional units called organs.

9–12 The Cell

★ Cells have particular structures that underlie their functions.

Related Ideas in *Benchmarks for Science Literacy* (AAAS 2009)

K–2 Cells

- Magnifiers help people see things they could not see without them.

3–5 Cells

★ Some living things consist of a single cell. Like familiar organisms, they need food, water, and air; a way to dispose of waste; and an environment they can live in.

★ Some organisms are made of a collection of similar cells that benefit from cooperating.

3–5 Diversity of Life

- There are millions of different kinds of individual organisms that inhabit the Earth at any one time—some very similar to each other, some very different.

6–8 Cells

★ All living things are composed of cells, from just one to many millions, whose details usually are visible only through a microscope.

- Different body tissues and organs are made up of different kinds of cells.
- Within cells, many of the basic functions of organisms—such as extracting energy from food and getting rid of waste—are carried out.
- The way in which cells function is similar in all living organisms.

6–8 Diversity of Life

- Animals and plants have a great variety of body plans and internal structures that contribute to their being able to make or find food and reproduce.

9–12 Cells

★ Within the cells are specialized parts for the transport of materials, energy capture and release, protein building, waste disposal, passing information, and even movement.

Related Research

- Studies by Dreyfus and Jungwirth (1988, 1989) of 16-year-old Israeli students revealed confusion about levels of organization in living things, including the idea that single-celled organisms contained organs such as intestines and lungs, even though they had been taught about cells in previous years (Driver at al. 1994).

★ Indicates a strong match between the ideas elicited by the probe and a national standard's learning goal.

Suggestions for Instruction and Assessment

- After students have responded to the probe, share the historical account of Leeuwenhoek's discovery of single-celled pond organisms he observed through one of the early microscopes he developed. Some students' ideas may be very similar to early ideas about these organisms. Leeuwenhoek called these organisms "animalcules" and thought they were like tiny animals, with details of their cell structures unknown at that time.
- Students should have opportunities to observe live, single-celled organisms. Collecting pond water will provide a variety of protists for students to observe. Culturing pond, lake, stream, or puddle water by making a hay infusion will increase the density of organisms for viewing. Directions for making hay infusions can be found on the internet. However, you should point out that small multicellular organisms, such as rotifers, do exist and could be mistaken for single-celled organisms.
- Probe deeper to find out what organs students think single-celled animals contain. Listen carefully for evidence of findings from a research study that shows some students think microorganisms contain digestive organs and respiratory organs such as lungs.
- Combine this probe with an opportunity for students to draw structures they think might be inside a single-celled organism that help it carry out its life processes.
- Consider using this probe to examine students' ideas about hierarchical organization. Encourage them to use logic to deduce that single-celled organisms could not contain organs because organs fall at a higher hierarchical level than cells.
- This probe context can be extended at the high school level by developing a similar probe to determine whether students think the internal structure of a bacterial cell (prokaryotic cell) is similar to the internal structure of a single-celled, eukaryotic organism. This more advanced probe can determine whether they recognize that some single-celled organisms have membrane-bound organelles, and others do not.

Related NSTA Science Store Publications, NSTA Journal Articles, NSTA SciGuides, NSTA SciPacks, and NSTA Science Objects

Anderson, R., and M. Druger. 2000. *Explore the world using protozoa.* Arlington, VA: NSTA Press.

Science Object: *Cell Structure and Function: Cells—The Basis of Life*

Related Curriculum Topic Study Guides (in Keeley 2005)
"Cells"
"Fungi and Microorganisms"

References

American Association for the Advancement of Science (AAAS). 2009. Benchmarks for science literacy online. *www.project2061.org/publications/bsl/online*

Dreyfus, A., and E. Jungwirth. 1988. The cell concept of 10th graders: Curricular expectations and reality. *International Journal of Science Education* 10 (2): 221–229.

Dreyfus, A., and E. Jungwirth. 1989. The pupil and the living cell: A taxonomy of dysfunctional ideas about an abstract idea. *Journal of Biological Education* 23 (1): 49–55.

Driver, R., A. Squires, P. Rushworth, and V. Wood-Robinson. 1994. *Making sense of secondary science: Research into children's ideas.* London: RoutledgeFalmer.

Keeley, P. 2005. *Science curriculum topic study: Bridging the gap between standards and practice.*

Thousand Oaks, CA: Corwin Press and Arlington, VA: NSTA Press.

National Research Council (NRC). 1996. *National science education standards*. Washington, DC: National Academies Press.

Atoms and Cells

The chart below shows a variety of things sorted into two different groups.

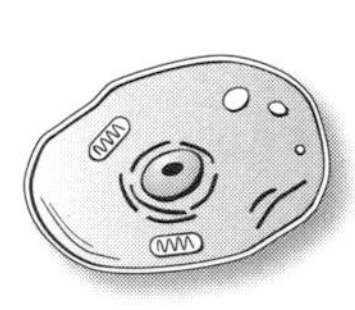

Group A	Group B
leaf of a plant	spoonful of salt
horse's muscle	piece of metal
cap of a mushroom	diamond necklace
baby elephant	protein
seed of a bean plant	sugar cube
blood	air

Circle the statement you think best describes the two groups.

A Both groups are made up of cells.

B Both groups are made up of atoms.

C Group A is made up of cells; Group B is made up of atoms.

D Group A is made up of cells and atoms; Group B is made up of atoms.

E Some things in Group A are made up of cells and some things in group B are made up of atoms.

F Group A and Group B are made up of both cells and atoms.

Explain your thinking. Provide an explanation for your answer.

____

____

____

____

____

Atoms and Cells

Teacher Notes

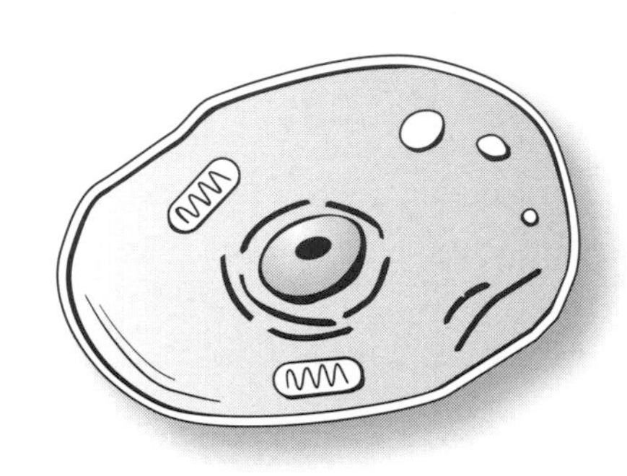

Purpose

The purpose of this assessment probe is to elicit students' ideas about the smallest parts of living and nonliving things. The probe is designed to determine how students distinguish between cells and atoms in living and nonliving contexts.

Related Concepts

Cells, atom, living

Explanation

The best answer is: D—Group A is made up of cells and atoms; Group B is made up of atoms. (*Note:* Students who choose answer B—Both groups are made of atoms—are correct. However, that would not be the *best* way to describe the two groups in order to differentiate between them.) All of the things in Group A are of a biological origin; that is, they are made up of cells. Cells are not the smallest unit of matter. They can be further broken down into molecules, which are made up of atoms. Except for the protein and sugar cube in Group B, all these things have a nonbiological origin and are also made up of atoms. The sugar in the cube and the protein were made by a living thing. Sugars and protein are made within a cell, but they are not made of cells. They are biomolecules, which can be further broken down into atoms.

Curricular and Instructional Considerations

Elementary Students

In the early grades, students explore parts and wholes and discover that objects and materials are made up of parts. They use magnifiers to examine small parts that are not obvious to the naked eye. At the upper elementary level, they are just beginning to learn that living things are made up of cells. They use simple microscopes to observe cells in familiar plant and animal parts as well as single-celled organisms. They begin to form a particle model of

matter, but details about atoms should wait until middle school.

Middle School Students

In middle school, students deepen their understanding of cells and recognize that cells are not only the basic unit of structure in living things but are also the basic unit of function. They develop a hierarchy to describe organization in living things from cell to tissue to organ to organ system to organism. They begin to use a particulate model of matter to explain various phenomena as well as to organize matter from its smallest unit, the atom, to molecules made up of atoms, to substances made up of molecules or arrays of atoms. By middle school, they should know that all living and nonliving matter is made up of atoms, but only living or once-living matter is made up of cells.

High School Students

Students' knowledge of cells has deepened to include cell organelles and biomolecules. Their knowledge of atoms now includes parts of the atom and interactions between atoms and between molecules. At this level, once students acquire a deeper understanding of atoms and molecules and of cellular structure and function, their atomic and molecular knowledge in chemistry and biology converge.

Administering the Probe

For high school students, consider adding more examples to Group A and B. For example, add DNA or a cell organelle such as a chloroplast to Group B, as some students may confuse DNA and cell organelles with cells if they lack an understanding of hierarchical structure. If this probe is used with upper elementary students, consider replacing unfamiliar items with familiar items.

Related Ideas in *National Science Education Standards* (NRC 1996)

K–4 Properties of Objects and Materials

- Objects are made up of different kinds of materials.

5–8 Structure and Function in Living Systems

- ★ All organisms are composed of cells, the fundamental units of life.
- ★ Important levels of organization for structure and function include cells, organs, tissues, organ systems, whole organisms, and ecosystems.
- Specialized cells perform specialized functions in multicellular organisms. Groups of specialized cells cooperate to form a tissue, such as a muscle. Different tissues are in turn grouped together to form larger functional units, called organs.

5–8 Structure and Function in Living Systems

- There are more than 100 elements that combine in a multitude of ways to produce compounds, which account for the living and nonliving substances that we encounter.

9–12 The Cell

- Cells have particular structures that underlie their functions. Every cell is surrounded by a membrane that separates it from the outside world. Inside the cell is a concentrated mixture of thousands of different molecules that form a variety of specialized structures.

9–12 Structure of Matter

- ★ Matter is made of minute particles called atoms, and atoms are composed of even smaller components.

★ Indicates a strong match between the ideas elicited by the probe and a national standard's learning goal.

★ Atoms often join with one another in various combinations in distinct molecules or in repeating three-dimensional crystal patterns.

Related Ideas in *Benchmarks for Science Literacy* (AAAS 2009)

K–2 Systems

- Most things are made of parts.

3–5 Cells

★ Microscopes make it possible to see that living things are made mostly of cells.

- Some organisms are made of a collection of similar cells that benefit from cooperating. Some organisms' cells vary greatly in appearance and perform very different roles in the organism.

3–5 Structure of Matter

- Materials may be composed of parts that are too small to be seen without magnification.

6–8 Cells

★ All living things are composed of cells, from just one to many millions, whose details usually are visible only through a microscope.

★ Different body tissues and organs are made up of different kinds of cells.

6–8 Structure of Matter

★ All matter is made up of atoms, which are far too small to see directly through a microscope.

- Atoms may link together in well-defined molecules, or may be packed together in crystal patterns. Different arrangements of atoms into groups compose all substances and determine the characteristic properties of substances.
- Carbon and hydrogen are common elements of living matter.

9–12 Cells

- The work of the cell is carried out by the many different types of molecules it assembles, mostly proteins.

9–12 Structure of Matter

- Atoms often join with one another in various combinations in distinct molecules or in repeating three-dimensional crystal patterns.
- An enormous variety of biological, chemical, and physical phenomena can be explained by changes in the arrangement and motion of atoms and molecules.

Related Research

- Understanding how molecules make up other small objects such as cells may be tied to the difficulty students have understanding how small something can be (Driver et al. 1994).
- Research conducted by Arnold (1983) indicated that students have difficulty differentiating between the concepts of *cell* and *molecule*. There is a tendency for students to overapply the idea that cells are smaller components of *living* things. Students identified any materials encountered in a biology class (carbohydrates, proteins, and water) as being made up of smaller parts called cells. Arnold coined the term *molecell* to describe the notion of organic molecules being considered as cells.
- Although many students eventually construct the idea that an atom or molecule is the smallest structural unit of a substance, they often have difficulty appreciating the minuteness of atoms and molecules (Driver et al. 1994).
- Students are unsure about the hierarchy of atoms, molecules, and cells. Cells are

★ Indicates a strong match between the ideas elicited by the probe and a national standard's learning goal.

described as the components of many things, including carbohydrates and proteins (Berthelsen 1999).

Suggestions for Instruction and Assessment

- Combine this probe with "Is It Made of Cells?" and "Is It Made of Molecules?" in *Uncovering Student Ideas in Science, Vol. 1: 25 Formative Assessment Probes* (Keeley, Eberle, and Farrin 2005).
- Use graphic organizers, such as Venn diagrams, to sort objects into things that are made of cells, things that are made of atoms, and things that are made of both cells and atoms.
- This probe can be turned into a FACT (formative assessment classroom technique) called a card sort (Keeley 2008). Instead of providing the chart on page 39, write each of the items listed on the chart on a card. Ask students to sort the cards into three groups: those items made up of cells, those items made up of atoms, and those items made up of cells and atoms. You may even add a fourth category—those made up of neither cells nor atoms (although no items go in this category, students may think otherwise). As students work in small groups to sort the cards, they must justify their reasoning for why they place each card in a category. As you circulate among the class, note how the cards are categorized and listen carefully to students' reasoning. Probe with further questions as needed. Use the information from this activity to inform a lesson on parts and wholes that addresses the students' misconceptions.
- To practice and learn the idea that big things are made of smaller parts, or for students to comprehend the relative sizes of things they learn about in one lesson compared to things in a new lesson, write a variety of terms on card stock (one item per card) and hand them out to the students in the class. Start with two random students and ask them to arrange themselves in size order (who is bigger). Then another student joins them and they all discuss "Where do I go?" Listen for various ways of reasoning, such as "I'm smaller than you because you are made up of me." For high school students, the cards might include items such as eukaryotic cell, bacterium, atom, electron, water molecule, glucose, seeds, cell nucleus, DNA, nucleotide, pinecone, ATP, protein, cat, nucleus, and mitochondrion.
- There are a number of video demonstrations that can help students understand the relative hierarchy of atoms, molecules, and cells. The best known example, although it may be a bit dated, is the Charles and Ray Eames *Powers of Ten* movie at *www.powersoften.com*.

Related NSTA Science Store Publications, NSTA Journal Articles, NSTA SciGuides, NSTA SciPacks, and NSTA Science Objects

American Association for the Advancement of Science (AAAS). 2001. *Atlas of science literacy.* Vol. 1. (See "Cells and Organs" map, pp. 74–75, and "Atoms and Molecules" map, pp. 54–55.) Washington, DC: AAAS.

Jones, G., A. Taylor, and M. Falvo. 2009. *Extreme science: From nano to galactic.* Arlington, VA: NSTA Press.

Minogue, J., G. Jones, B. Broadwell, and T. Oppewall. 2006. Exploring cells from the inside out: New tools for the classroom. *Science Scope* 29 (6): 28–32.

Science Object: *Cell Structure and Function: Cells—The Basis of Life*

Science Object: *Cell Structure and Function: The Molecular Machinery of Life*

SciPack: *Atomic Structure*

Related Curriculum Topic Study Guides (in Keeley 2005)
"Cells"
"Particulate Nature of Matter (Atoms and Molecules)"

References

American Association for the Advancement of Science (AAAS). 2009. Benchmarks for science literacy online. *www.project2061.org/publications/bsl/online*

Arnold, B. 1983. Beware the molecell! *Biology Newsletter* (Aberdeen College of Education) 42: 2–6.

Berthelsen, B. 1999. Students' naive conceptions in life science. *Michigan Science Teachers Association Journal* 44 (1): 13–19. *www.msta-mich.org*

Driver, R., A. Squires, P. Rushworth, and V. Wood-Robinson. 1994. *Making sense of secondary science: Research into children's ideas.* London: RoutledgeFalmer.

Keeley, P. 2005. *Science curriculum topic study: Bridging the gap between standards and practice.* Thousand Oaks, CA: Corwin Press and Arlington, VA: NSTA Press.

Keeley, P. 2008. *Science formative assessment: 75 practical strategies for linking assessment, instruction, and learning.* Thousand Oaks, CA: Corwin Press and Atlington, VA: NSTA Press.

Keeley, P., F. Eberle, and L. Farrin. 2005. *Uncovering student ideas in science, vol. 1: 25 formative assessment probes.* Arlington, VA: NSTA Press.

National Research Council (NRC). 1996. *National science education standards.* Washington, DC: National Academies Press.

Which One Will Dry Out Last?

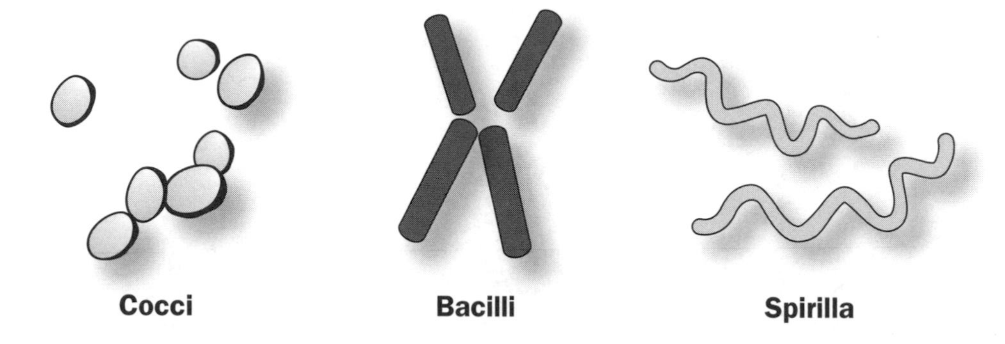

Bacterial cells have three common shapes—cocci, bacilli, and spirilla. It is important for bacteria to have a moist environment so they do not dry out. Under hot, dry conditions, which of the bacterial cell shapes is most likely to dry out last if each cell has the same volume? Circle your answer.

A Cocci

B Baccilli

C Spirilla

D They would all dry out at the same time.

Explain your thinking. What "rule" or reason did you use to select your answer?

____

____

____

____

____

Which One Will Dry Out Last?

Teacher Notes

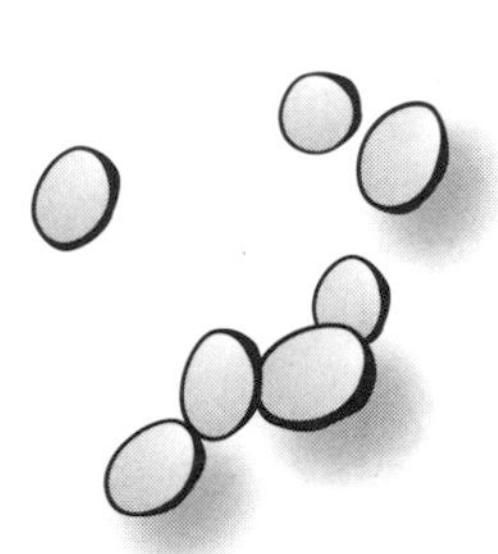

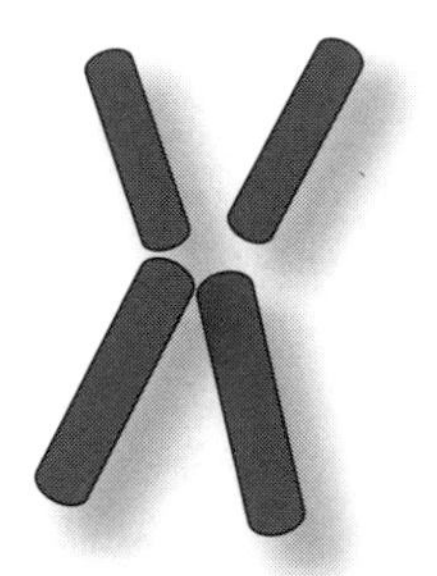

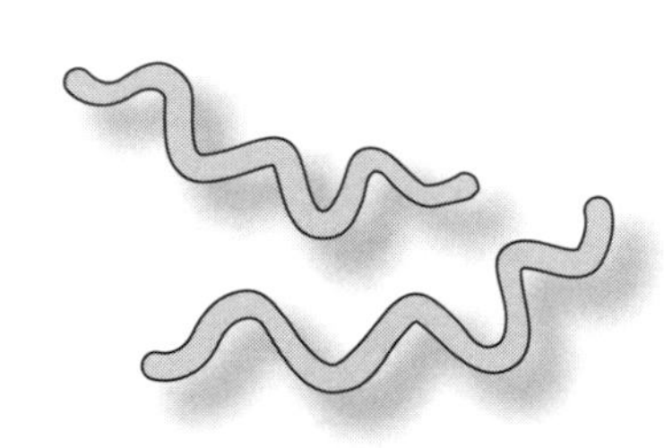

Purpose

The purpose of this assessment probe is to elicit students' ideas about loss of water through cells. The probe is designed to reveal whether students recognize the relationship between shape, surface area, and volume.

Related Concepts

Cells, cell shape, bacteria, single-celled organisms, volume, surface area

Explanation

The best answer is A—cocci. The answer is best explained through geometry. Cocci, spherical-shaped bacteria, have the lowest surface area to volume ratio. Given the same volume, spheres have the smallest surface area of any shape. The low ratio of surface area to volume results in less water loss through the cell. The ratio of surface area to volume is important to cells because more contact with the environment through the surface of a cell relative to its environment increases water loss. Long, thinner cells, like the spirilla, have a greater ratio of surface area to volume and therefore would dry out first.

However, bear in mind this is the "best" answer, based on general shape, but shape alone cannot be used to predict the rate of drying for some bacteria. Bacteria have a variety of physical and biochemical defenses against desiccation. For example, two bacilli of the same size and shape can have very different rates of drying. Moreover, many bacteria can form very resistant spores that defy predictions based on shape alone.

Curricular and Instructional Considerations

Elementary Students

In the elementary grades, students observe a variety of plant and animal cells and describe their shapes. They learn about the need that all living things have for water. In mathematics they learn about and distinguish between a variety of three-dimensional shapes.

Middle School Students

Middle school students deepen their knowledge of cells, including single-celled organisms. They learn that cells are mostly made of water and that cells must control the passage of water in and outside of the cell. They learn about the basic characteristics of bacteria and develop a notion that bacteria are much smaller than the familiar cells they have observed. In mathematics, they develop an understanding of surface area and volume and apply this to several different phenomena in science. For example, water will evaporate more quickly from an open container that has an opening with a large surface area than from a container of the same volume with a smaller opening. They also deepen their knowledge of ratios at this level and use ratios to describe relationships.

High School Students

High school students deepen their understanding of the factors that affect the passage of water in and out of a cell. They learn about the cell membrane in more detail. They explore the ratio of surface area to volume in cells, organs, and organisms and use it to explain several important biological ideas. Using mathematics, they can compare surface areas of different shapes and predict what happens when the surface areas increase while the volume stays the same.

Administering the Probe

This probe is most appropriate for high school students. Show students microscopic photos of the three different shapes of bacteria. Please note that it is difficult to illustrate volume in a two-dimensional drawing. Make sure students know that the volume is the same for all three cells in this example by explicitly stating this to the students. It is the shape that changes.

Related Ideas in *National Science Education Standards* (NRC 1996)

K–4 The Characteristics of Organisms

- Organisms have basic needs. For example, animals need air, water, and food; plants require air, water, nutrients, and light. Organisms can survive only in environments where their needs can be met.

5–8 Structure and Function in Living Systems

- All organisms are composed of cells—the fundamental unit of life. Most organisms are single cells.
- Cells carry on the many functions needed to sustain life.

9–12 The Cell

- Cells have particular structures that underlie their functions. Every cell is surrounded by a membrane that separates it from the outside world.

Related Ideas in *Benchmarks for Science Literacy* (AAAS 2009)

K–2 The Cell

- Most living things need water, food, and air.

3–5 The Cell

- Some living things consist of a single cell. Like familiar organisms, they need food, water, and air; a way to dispose of waste; and an environment they can live in.

6–8 Scale

- Some properties of an object depend on its length, some depend on its area, and some depend on its volume.

6–8 The Cell

- All living things are composed of cells, from just one to many millions, whose details usually are visible only through a microscope.
- About two-thirds of the weight of cells is accounted for by water, which gives cells many of their properties.

9–12 Scale

★ Because different properties are not affected to the same degree by changes in size, large changes in size typically change the way that things work in physical, biological, or social systems.

9–12 The Cell

- Every cell is covered by a membrane that controls what can enter and leave the cell.

Related Research

- This question is similar to one used by Livne (1996) in which high school biology students were asked about the cooling of milk. They were given a spherical baby bottle and a cylindrical baby bottle and told that they each contained the same amount of milk. They were then asked which bottle of milk would cool off fastest, or whether they would cool at the same time. Fewer than 50% predicted the cylindrical bottle would cool faster because of the greater surface area. Most predicted they would cool at the same rate because their volumes were equal. The students were then presented with the question about which type of bacteria are most resistant to drying. About 40% recognized the ratio of surface area to volume and correctly predicted the spherical bacteria (cocci) would be most resistant. However, the majority of students said they would dry out at the same time because they had the same volume.
- Stavy and Tirosh (2000) identify the intuitive rule "same A, same B" to explain why students choose that the bacteria would dry out at the same time. Their reasoning is that if the bacteria have the same volume, then they have the same resistance to drying.

Suggestions for Instruction and Assessment

- To further address the relationship between surface area and volume and issues related to scale, consider combining this probe with "Whale and Shrew" in *Uncovering Student Ideas in Science, Vol. 2: 25 More Formative Assessment Probes* (Keeley, Eberle, and Tugel 2007).
- After students have made their predictions and explained their answers, show 3-D models that have nearly identical volumes and approximately the shapes of the bacteria—for example, a Styrofoam ball, a wooden rod, and a piece of thick rope. Ask students to imagine these are the bacterial cells and that they have the same volume. Ask if they would change their predictions after looking at the models. Alternatively, provide students with modeling clay. Have them create the three shapes from the same volume of clay, then calculate and compare the surface areas of each shape.
- For high school students, connect the surface area to volume relationship to the idea of cells needing to diffuse material quickly in and out for gas exchange, food, and so on, in addition to maintaining water content. Because most cells normally exist in an aqueous environment, water loss is less of an issue than the ongoing need to bring in materials for food or to exchange gases. These needs are also related to the relationship between surface area and volume.
- Give students a ball of clay and have them mathematically calculate the relationship

★ Indicates a strong match between the ideas elicited by the probe and a national standard's learning goal.

between surface area and volume. Have them shape the clay into a rectangular solid and perform the same calculation. Then have them reshape the clay into a long ropelike cylinder shape and perform the calculation again. Have them compare all three of the volumes (they are the same), the surface areas, and the ratio of surface area to volume. Have them compare their results to the three bacterial cell shapes in the probe.

- As a postassessment, this probe can be modified to ask students to predict and explain the order in which these cells would dry out if exposed to environmental conditions that promote desiccation.
- Challenge students to come up with other examples of when a sphere is the best shape for a particular circumstance. Contrast this with examples of when it is better to have a large exposed surface area.

Related NSTA Science Store Publications, NSTA Journal Articles, NSTA SciGuides, NSTA SciPacks, and NSTA Science Objects

American Association for the Advancement of Science (AAAS). 2001. *Atlas of science literacy.* Vol. 1. (See "Cells and Organs" map, pp. 74–75.) Washington, DC: AAAS.

Konicek-Moran, R. 2008. Dried apples. In *Everyday science mysteries: Stories for inquiry-based science teaching,* 99–106. Arlington, VA: NSTA Press.

Maguire, L., L. Myerowitz, and V. Sampson. 2010. Exploring osmosis and diffusion in cells. *The Science Teacher* 77 (8): 55–60.

SciPack: *Cell Structure and Function*

Related Curriculum Topic Study Guides (in Keeley 2005)
"Cells"
"Scale"

References

American Association for the Advancement of Science (AAAS). 2009. Benchmarks for science literacy online. *www.project2061.org/publications/bsl/online.*

Keeley, P. 2005. *Science curriculum topic study: Bridging the gap between standards and practice.* Thousand Oaks, CA: Corwin Press and Arlington, VA: NSTA Press.

Keeley, P., F. Eberle, and J. Tugel. 2007. *Uncovering student ideas in science, vol. 2: 25 more formative assessment probes.* Arlington, VA: NSTA Press.

Livne, T. 1996. Examination of high school students' difficulties in understanding the change in surface area, volume, and surface area/volume ratio with the change in size and/or shape of a body. Unpublished master's thesis, Tel Aviv University, Tel Aviv, Israel.

National Research Council (NRC). 1996. *National science education standards.* Washington, DC: National Academies Press.

Stavy, R., and D. Tirosh. 2000. *How students (mis-) understand science and mathematics: Intuitive rules.* New York: Teachers College Press.

Chlorophyll

Chlorophyll is a substance found in the cells of plants. Put an X next to all the things you think are a function of *chlorophyll.*

A ______ provides protection for the leaf

B ______ controls the odors that come from a plant

C ______ serves as a storage product in a plant

D ______ is a source of food for a plant

E ______ absorbs light energy

F ______ absorbs carbon dioxide

G ______ strengthens plant cells

H ______ breaks down sugars and starches

I ______ gives most plants their green color

J ______ is a vital liquid in plants that acts similar to blood in animals

K ______ attracts bees and butterflies to the plant for pollination

L ______ manufactures food for the plant

Explain your thinking. Describe what you know about *chlorophyll.*

Chlorophyll

Teacher Notes

Purpose

The purpose of this assessment probe is to elicit students' ideas about a scientific word they frequently encounter in middle and high school science, *chlorophyll.* The probe is designed to reveal whether students know that chlorophyll is more than just a pigment. Their answers reveal whether they also recognize that its function is to absorb light energy for photosynthesis.

Related Concepts

Photosynthesis, chlorophyll, chloroplast, cells, leaf, energy, plants

Explanation

There are two best responses—E and I. Chlorophyll is a pigment found in the chloroplast of plant cells (as well as many single-celled organisms). The primary function of chlorophyll is to absorb light energy (E). Chloroplasts convert light energy to chemical energy using chlorophyll. Chlorophyll also makes most plants appear green (I), although this is a passive function as the chlorophyll molecule simply reflects green light rather than absorbs it and makes plants appear green to our eyes. However, its main purpose is not to provide color for a plant but rather to act as a photoreceptor. Chlorophyll is only one part of the multicomponent organelle, chloroplast. The unique structure of the chlorophyll molecule allows it to function in absorbing particles of light energy (photons) from sunlight and then use them to excite electrons. These electrons are passed to other molecules in a large, complex reaction center. The other molecules in the reaction center convert the energy from sunlight to chemical energy, which is then stored temporarily in the first part of photosynthesis. This energy is then used when the carbon in CO_2 is fixed into glucose during the second part of photosynthesis.

Curricular and Instructional Considerations

Elementary Students

In the elementary grades, students wonder about the differences between plants and animals and ask questions such as "How do plants get food?" (NRC 1996, p. 128). They learn that plants need nutrients and are introduced to the idea that plants make their own food, rather than acquire it like animals, but concepts related to the details of photosynthesis are not developed until middle school.

Middle School Students

In middle school, students are introduced to the basic process of photosynthesis and make connections between the matter and energy ideas that are part of understanding this vital plant process. They learn about plant cells and how they differ from animal cells. Although details related to cell organelles can wait until high school, students are introduced to *chloroplasts* and the pigment, *chlorophyll.* These "terms" are connected to their understanding of the leaf as the plant structure that takes in carbon dioxide and absorbs sunlight for energy used in the process of photosynthesis. They examine leaves microscopically and observe the green chloroplasts within the leaf cells that carry out the process of photosynthesis.

High School Students

At the high school level, students deepen their understanding of photosynthesis, including the molecules involved. In high school, they should be familiar with the word *chlorophyll,* and know that it is a pigment found in the chloroplast that absorbs light energy, which the chloroplast converts to chemical energy. They learn that there are two types of chlorophyll in plants: chlorophyll a and chlorophyll b. They also connect their understanding of the visible spectrum to the wavelengths of light absorbed by the chlorophyll molecule to understand why plants appear green.

Administering the Probe

This probe is best used at the middle and high school levels. It should be used only if students have encountered the word *chlorophyll* through previous experiences or instruction. This probe can also be administered as a card sort, one of the FACTs (formative assessment classroom techniques) used with probes in the form of a list (Keeley 2008). Write each choice listed on page 51 on a separate card. Make four or five sets of the cards. Have students work in pairs or small groups to sort the cards into those that describe chlorophyll and those that do not. Remind students to explain their reasoning as they sort each card. As you circulate among the class, notice where the cards are placed, listen to students explain their reasoning, and probe further to find out what they think they know about chlorophyll.

Related Ideas in *National Science Education Standards* (NRC 1996)

K–4 The Characteristics of Organisms

- Organisms have basic needs. For example, animals need air, water, and food; plants require air, water, nutrients, and light.
- Each plant or animal has different structures that serve different functions in growth, survival, and reproduction.

5–8 Structure and Function in Living Systems

- Specialized cells perform specialized functions in multicellular organisms. Groups of specialized cells cooperate to form a tissue.

5–8 Populations and Ecosystems

- Plants and some microorganisms are producers—they make their own food.

- ★ For ecosystems, the major source of energy is sunlight. Energy entering ecosystems as sunlight is transferred by producers into chemical energy by photosynthesis.

9–12 The Cell

- ★ Plant cells contain chloroplasts, the site of photosynthesis. Plants and many microorganisms use solar energy to combine molecules of carbon dioxide and water into complex, energy-rich organic compounds and release oxygen to the environment. This process of photosynthesis provides a vital connection between the Sun and the energy needs of living systems.

Related Ideas in *Benchmarks for Science Literacy* (AAAS 2009)

K–2 Flow of Matter and Energy in Ecosystems

- Plants and animals both need to take in water, and animals need to take in food. In addition, plants need light.

K–2 Cells

- Most living things need water, food, and air.

3–5 Cells

- Some organisms are made of a collection of similar cells that benefit from cooperating. Some organisms' cells vary greatly in appearance and perform very different roles in the organism.

3–5 Flow of Matter and Energy in Ecosystems

- Some source of "energy" is needed for all organisms to stay alive and grow.

6–8 Cells

- Various organs and tissues function to serve the needs of all cells for food, air, and waste removal.

6–8 Flow of Matter and Energy in Ecosystems

- Food provides molecules that serve as fuel and building material for all organisms.
- ★ Plants use the energy from light to make sugars from carbon dioxide and water.
- Energy can change from one form to another in living things.
- Almost all food energy originally comes from sunlight.

9–12 Cells

- ★ Within the cells are specialized parts for the transport of materials, energy capture and release, protein building, waste disposal, passing information, and even movement.
- In addition to the basic cellular functions common to all cells, most cells in multicellular organisms perform some special functions that others do not.

Related Research

- Children appear to think of chlorophyll in one of six ways: as a food substance, a protection, a storage product, a vital substance like blood, something that makes plants strong, or something that breaks down starch (Driver et al. 1994).
- Children who have some notion of its function in photosynthesis think that chlorophyll either attracts sunlight or absorbs carbon dioxide (Driver et al. 1994).
- Some children think chlorophyll's primary purpose is to make leaves green and attractive (Bell and Brook 1984).
- Barker (1985) found that even after students have been taught the role of chlorophyll in absorbing light energy, the students rarely learned that concept.

★ Indicates a strong match between the ideas elicited by the probe and a national standard's learning goal.

- In Simpson and Arnold's study (1982), only 29% of 12- to 13-year-olds, and 46% of 14- to 16-year-olds, understood the role of chlorophyll in converting light energy to chemical energy during photosynthesis.

Suggestions for Instruction and Assessment

- Combine this probe with "Giant Sequoia Tree" in *Uncovering Student Ideas in Science, Vol. 2: 25 More Formative Assessment Probes* (Keeley, Eberle, and Tugel 2007), which uses chlorophyll as a distracter for where most of the mass of a tree comes from.
- Students should have opportunities to learn the concept that plant cells have a pigment that absorbs light energy before the teacher formally introduces the vocabulary word *chlorophyll.* Learning the concept first gives the students an idea to hang the word on to.
- Combine this probe with another FACT called a Scientific Terminology Inventory Probe (STIP) (Keeley 2008). STIP allows teachers to assess students' prior knowledge of and familiarity with the *technical terms* they will encounter in a unit. STIP is a list of terms related to an upcoming unit. Next to each term, students check off whether they have never heard the term before, have heard the term before but are not sure what it means, have some idea of what it means, or clearly know what it means and can explain it.
- Combine an understanding of the chlorophyll pigment with how we see the color green when we look at plants. Use a prism to show how light breaks down into a visible spectrum. Challenge students to describe which wavelengths and colors are absorbed by chlorophyll and which ones are reflected. This activity provides a connection between biological and physical science concepts once students have been introduced to the visible spectrum.

Related NSTA Science Store Publications, NSTA Journal Articles, NSTA SciGuides, NSTA SciPacks, and NSTA Science Objects

American Association for the Advancement of Science (AAAS). 2001. *Atlas of science literacy.* Vol. 1. (See "Flow of Matter in Ecosystems" map, pp. 76–77 and "Flow of Energy in Ecosystems" map, pp. 78–79.) Washington, DC: AAAS.

Keeley, P. 2008. *Science formative assessment: 75 practical strategies for linking assessment, instruction, and learning.* Thousand Oaks, CA: Corwin Press and Arlington, VA: NSTA Press.

Koba, S., with A. Tweed. 2009. *Hard-to-teach biology concepts: A framework to deepen student understanding.* (See Chapter 4, "Flow of Energy and Matter: Photosynthesis," pp. 119–142.) Arlington, VA: NSTA Press.

Konicek-Moran, R. Forthcoming. Lichen lookin'. In *Yet more everyday science mysteries: Stories for inquiry-based science learning.* Arlington, VA: NSTA Press.

Littlejohn, P. 2007. Building leaves and an understanding of photosynthesis. *Science Scope* 30 (8): 22–25.

Mundry, S., P. Keeley, and C. Landel. 2009. *A leader's guide to science curriculum topic study.* (See Module B6, Photosynthesis and Respiration Facilitation Guide, pp. 144–149.) Thousand Oaks, CA: Corwin Press.

Robertson, W. 2007. Science 101: How does photosynthesis work? *Science and Children* 44 (8): 60–63.

Related Curriculum Topic Study Guides (in Keeley 2005)
"Cells"
"Photosynthesis and Respiration"

References

American Association for the Advancement of Science (AAAS). 2009. Benchmarks for science literacy online. *www.project2061.org/publications/bsl/online*

Barker, M. 1985. *Teaching and learning about photosynthesis.* Science Education Research Unit, Working Papers 220–229, University of Waikato, Hamilton, New Zealand.

Bell, B., and A. Brook. 1984. *Aspects of secondary students' understanding of plant nutrition.* Leeds, UK: University of Leeds, Children's Learning in Science Project, Centre for Studies in Science and Mathematics Education.

Driver, R., A. Squires, P. Rushworth, and V. Wood-Robinson. 1994. *Making sense of secondary science: Research into children's ideas.* London: RoutledgeFalmer.

Keeley, P. 2005. *Science curriculum topic study: Bridging the gap between standards and practice.* Thousand Oaks, CA: Corwin Press and Arlington, VA: NSTA Press.

Keeley, P., F. Eberle, and J. Tugel. 2007. *Uncovering student ideas in science, vol 2: 25 more formative assessment probes.* Arlington, VA: NSTA Press.

National Research Council (NRC). 1996. *National science education standards.* Washington, DC: National Academies Press.

Simpson, M., and B. Arnold. 1982. Availability of prerequisite concepts for learning biology at certificate level. *Journal of Biological Education* 16 (1): 65–72.

Apple Tree

Six friends were picking apples. They each had different ideas about where the apple tree makes the food it needs to live and grow. This is what they said:

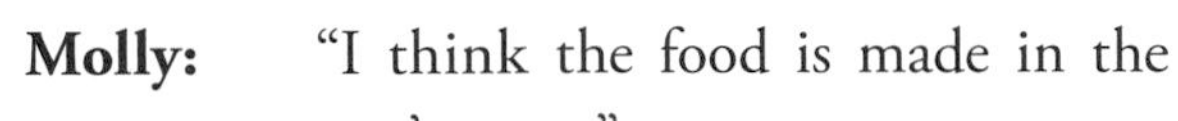

Molly: "I think the food is made in the tree's roots."

Joan: "I think the food is made in the leaves of the tree."

Bonnie: "I think the food is made in the apples the tree produces."

Bev: "I think the food is made in the tiny tubes in the trunk and the branches."

Susie: "I think the food is made in the apple blossoms."

Jared: "I disagree with all of you. I don't think apple trees make their own food."

Which friend do you agree with the most? ________ Explain why you agree.

____

____

____

____

____

____

____

____

____

____

Apple Tree

Teacher Notes

Purpose

The purpose of this assessment probe is to elicit students' ideas about the plant structure where photosynthesis takes place. The probe is designed to reveal whether students recognize that food is made within the leaf of a plant.

Related Concepts

Food, photosynthesis, leaf, plants, energy, chlorophyll, chloroplast

Explanation

The best response is Joan's: "I think the food is made in the leaves of the tree." Trees are plants and, therefore, they make their own food. Plants take in carbon dioxide through their leaves and water through their roots. The chloroplasts in the leaves use energy from sunlight to break apart and reassemble these inorganic molecules into an organic molecule in the form of a simple sugar (glucose), giving off oxygen as a by-product. This process is called photosynthesis and happens inside the chloroplast of a plant cell. Photosynthesis happens primarily in the leaves of plants. You can think of a leaf as a solar collector full of photosynthetic cells. Solar energy strikes the leaf and is captured by the chlorophyll inside the leaf cells' chloroplasts. The carbon dioxide for photosynthesis enters the leaves from the air, while the roots take up the water that is needed and the xylem transports it to parts of the plant, including the leaves.

Using energy from sunlight absorbed by chlorophyll, these materials are transformed into food (sugar) that is made in the leaves. These sugars travel through vessels in the veins of leaves that transport the food throughout the plant. Oxygen, a by-product of the photosynthetic process, passes out of the leaf into the air. Some of the sugar is also used to make amino acids.

Although this is a simplified description of a complex process, the important point is that the leaf is the structure in plants that functions in transforming materials from the

environment into food (glucose), which is used as building blocks for the plant.

Curricular and Instructional Considerations

Elementary Students

In the elementary grades, students learn that plants need water, sunlight, nutrients, and air. They learn that plants have structures for getting water (roots) and sunlight (leaves). They learn that stems provide support for the plant and flowers are used in reproduction. They are introduced to the idea that plants make their own food, but the details of the process should wait until middle school.

Middle School Students

In middle school, students are introduced to the basic process of photosynthesis and the plant structures involved in the process. They learn about the parts of the leaf and the cells involved in the food-making process. They begin to develop an understanding of the distinction between the role of sunlight as a source of energy for the process, the transformation of light energy to chemical energy, and the inorganic materials transformed during the process. They learn about the vascular tissue (xylem and phloem) within the plant that transports water from the roots and food made in the leaves to all parts of the plant. Middle school is also the time when students develop a scientific conception of *food* that differs from the common, everyday use of the word *food*.

High School Students

In high school, students deepen their understanding of photosynthesis by moving from an emphasis on what happens in the leaf to understanding what happens within the cell, especially at the molecular level. They connect their growing understanding of chemistry to the biological process of photosynthesis. They learn about the different types of carbohydrates formed by plants from simple sugar (glucose). They distinguish between the energy provided by sunlight and the energy released from the chemical bonds formed between atoms that make up the molecules of sugar.

Administering the Probe

This probe is best used at the upper elementary and middle level, before students learn about the structure and function of plants. It can also be used with high school students before they learn about photosynthesis to determine if their earlier misconceptions are still present.

Related Ideas in *National Science Education Standards* (NRC 1996)

K–4 The Characteristics of Organisms

- Organisms have basic needs. For example, animals need air, water, and food; plants require air, water, nutrients, and light.
- Each plant or animal has different structures that serve different functions in growth, survival, and reproduction.

5–8 Structure and Function in Living Systems

★ Specialized cells perform specialized functions in multicellular organisms. Groups of specialized cells cooperate to form a tissue.

5–8 Populations and Ecosystems

- Plants and some microorganisms are producers—they make their own food.
- For ecosystems, the major source of energy is sunlight. Energy entering ecosystems as sunlight is transferred by producers into chemical energy by photosynthesis.

9–12 The Cell

★ Plant cells contain chloroplasts, the site of photosynthesis. Plants and many micro-

★ Indicates a strong match between the ideas elicited by the probe and a national standard's learning goal.

organisms use solar energy to combine molecules of carbon dioxide and water into complex, energy-rich organic compounds and release oxygen to the environment. This process of photosynthesis provides a vital connection between the Sun and the energy needs of living systems.

Related Ideas in *Benchmarks for Science Literacy* (AAAS 2009)

K–2 Flow of Matter and Energy in Ecosystems

- Plants and animals both need to take in water, and animals need to take in food. In addition, plants need light.

K–2 Cells

- Most living things need water, food, and air.

3–5 Cells

- Some organisms are made of a collection of similar cells that benefit from cooperating. Some organisms' cells vary greatly in appearance and perform very different roles in the organism.

3–5 Flow of Matter and Energy in Ecosystems

- Some source of "energy" is needed for all organisms to stay alive and grow.

6–8 Cells

★ Various organs and tissues function to serve the needs of all cells for food, air, and waste removal.

6–8 Flow of Matter and Energy in Ecosystems

- Food provides molecules that serve as fuel and building material for all organisms.
- Plants use the energy from light to make sugars from carbon dioxide and water.
- Plants can use the food they make immediately or store it for later use.

9–12 Cells

★ Within the cells are specialized parts for the transport of materials, energy capture and release, protein building, waste disposal, passing information, and even movement.

- In addition to the basic cellular functions common to all cells, most cells in multicellular organisms perform some special functions that others do not.

Related Research

- Universally, the most persistent notion that students of all ages have about how plants get their food is that plants take their food from the environment, particularly the soil (Driver et al. 1994).
- In a study by Tamir (1989), some students thought sunlight, associated with energy, was the food for plants. Many students also considered minerals taken in from the soil as food.
- *Food* is colloquially understood to be anything an organism takes in for nourishment; therefore, students believe that anything absorbed from the soil is food. This is an instance where everyday meaning and scientific meaning clash and create confusion. Garden fertilizers labeled "plant food" reinforce this erroneous idea that fertilizer is food for plants (Allen 2010).
- The common misconception that plants get their food from the environment rather than manufacturing it internally, and that food for plants is taken in from the outside, is particularly resistant to change, even after instruction (Anderson, Sheldon, and Dubay 1990).
- Understanding that the food plants make is very different from nutrients they take in may be a prerequisite for understanding the idea that plants make their food rather

★ Indicates a strong match between the ideas elicited by the probe and a national standard's learning goal.

than acquire it from their environments (Roth, Smith, and Anderson 1983).
- Some students do not consider trees to be plants (Driver et al. 1994). Therefore, they may believe trees do not make their own food.

Suggestions for Instruction and Assessment

- Combine this probe with "Is It Food for Plants?" or "Is It a Plant?" in *Uncovering Student Ideas in Science, Vol. 2: 25 More Formative Assessment Probes* (Keeley, Eberle, and Tugel 2007). "Is It a Plant?" will help first determine whether students recognize trees as being plants.
- Take the time to elicit students' definitions of the word *food;* many students use this word in a way that is not consistent with its biological meaning (AAAS 2009). For example, many students think of food as something that is taken in through the mouth. Have students identify the difference between the everyday use of the word *food* and the scientific use of the word. Contrasting the two uses and providing examples may help students recognize the difference and know when the word is being used in a biological context.
- Some students confuse the sweet apple with the sugars that plants make during photosynthesis. Be sure to teach what happens to the sugars that plants produce in their leaves so that students recognize how the sugars are the building blocks for parts of the plant and are transported to other parts of the plant where they are used or stored for later use.
- Students may confuse the phloem (a structure that transports the sugars throughout a plant) with the site where the sugars are made. Be sure to distinguish between the function of the leaf in manufacturing food and the network of tiny tubes that transport the food to other parts of the plant.

Related NSTA Science Store Publications, NSTA Journal Articles, NSTA SciGuides, NSTA SciPacks, and NSTA Science Objects

American Association for the Advancement of Science (AAAS). 2001. *Atlas of science literacy.* Vol. 1. (See "Flow of Matter in Ecosystems" map, pp. 76–77 and "Flow of Energy in Ecosystems" map, pp. 78–79.) Washington, DC: AAAS.

Koba, S., with A. Tweed. 2009. *Hard-to-teach biology concepts: A framework to deepen student understanding.* (See Chapter 4, Photosynthesis and Respiration, pp. 119–141.) Arlington, VA: NSTA Press.

Littlejohn, P. 2007. Building leaves and an understanding of photosynthesis. *Science Scope* 30 (8): 22–25.

Mundry, S., P. Keeley, and C. Landel. 2009. *A leader's guide to science curriculum topic study.* (See Module B6, Photosynthesis and Respiration Facilitation Guide, pp. 144–149). Thousand Oaks, CA: Corwin Press.

Robertson, W. 2007. Science 101: How does photosynthesis work? *Science and Children* 44 (8): 60–63.

Wheeler-Toppen, J. 2010. The case of the tree hit man. In *Once upon a life science book: 12 interdisciplinary activities to create confident readers.* Arlington, VA: NSTA Press.

Related Curriculum Topic Study Guides (in Keeley 2005)
"Food and Nutrition"
"Photosynthesis and Respiration"
"Plant Life"

References

Allen, M. 2010. *Misconceptions in primary science.* Berkshire, UK: Open University Press.

American Association for the Advancement of Science (AAAS). 2009. Benchmarks for science

literacy online. *www.project2061.org/publications/bsl/online*

Anderson, C., T. Sheldon, and J. Dubay. 1990. The effects of instruction on college non-majors' conceptions of respiration and photosynthesis. *Journal of Research in Science Teaching* 27: 776–776.

Driver, R., A. Squires, P. Rushworth, and V. Wood-Robinson. 1994. *Making sense of secondary science: Research into children's ideas.* London: RoutledgeFalmer.

Keeley, P. 2005. *Science curriculum topic study: Bridging the gap between standards and practice.* Thousand Oaks, CA: Corwin Press and Arlington, VA: NSTA Press.

Keeley, P., F. Eberle, and J. Tugel. 2007. *Uncovering student ideas in science, vol. 2: 25 more formative assessment probes.* Arlington, VA: NSTA Press.

National Research Council (NRC). 1996. *National science education standards.* Washington, DC: National Academies Press.

Roth, K., E. Smith, and C. Anderson. 1983. *Students' conceptions of photosynthesis and food for plants.* East Lansing, MI: Michigan State University, Institute for Research on Teaching.

Tamir, P. 1989. Some issues related to justifications to multiple choice answers. *Journal of Biological Education* 11 (1): 48–56.

Light and Dark

Five friends were talking about when plants carry out the processes of photosynthesis and respiration. This is what they said:

Janet: "Photosynthesis and respiration occur both when it is light and when it is dark."

Calvin: "Photosynthesis occurs when it is light; respiration occurs when it is dark."

Mika: "Photosynthesis occurs when it is light; respiration occurs both when it is light and when it is dark."

Turner: "Photosynthesis occurs both when it is light and when it is dark; respiration happens at night."

Sophie: "Photosynthesis occurs in the light; plants don't carry out the process of respiration."

Whom do you agree with the most? ____________________ Explain why you agree with that person and not the others.

Light and Dark

Teacher Notes

Purpose

The purpose of this assessment probe is to elicit students' ideas about when the processes of photosynthesis and respiration occur. The probe is designed to reveal whether students recognize that plants respire continuously.

Related Concepts

Photosynthesis, respiration, energy, plants

Explanation

The best answer is Mika's: "Photosynthesis occurs when it is light; respiration occurs both when it is light and when it is dark." The word *photosynthesis* means to "make with light." Plants photosynthesize during the daytime when sunlight is available (or at other times when they are exposed to artificial light). They capture light energy from the Sun, which they convert into chemical energy during the process of photosynthesis. Plant cells, like all other cells, continuously need energy to carry out their life processes. Cellular respiration is the process during which cells take in oxygen to break down sugars and produce adenosine triphosphate (ATP) that can be used by the cell. This process does not require light. It requires a food molecule like glucose and oxygen (for aerobic respiration). This process (cellular respiration) occurs any time the cell needs energy to carry out its life processes, whether it is night or day. A common misconception is that plants photosynthesize during the day and conduct cellular respiration only at night. This misconception was even perpetuated when hospitals used to remove plants from a patient's room during the night to ensure an adequate oxygen supply for the patient!

Light is required to drive one of the two main photosynthetic reaction processes. This light-dependent reaction uses the energy from sunlight to produce high-energy electrons (which are stored in ATP and NADPH). This light-dependent reaction also produces oxygen (O_2) as a waste product. Simultaneously, a second set of reactions called the Calvin

cycle uses the ATP and NADPH that are formed during the light reaction to produce high-energy sugars. The Calvin cycle occurs at the same time as the light-dependent reactions inside the chloroplast. The reactions in the Calvin cycle can happen in the light or in the darkness, as they are not light dependent. This is why the reactions in the cycle also may be called the light-independent or dark reactions. However, the process of carbon fixation is light-dependent overall, because without the energy from light the chloroplasts will deplete the ATP and NADPH required for the reaction, and the Calvin cycle will stop. Students who recognize that there is a light and dark reaction may believe erroneously that the dark reaction can only occur in darkness.

Photosynthesis and cellular respiration occur in different sites within the cell. Photosynthesis occurs in the chloroplast and respiration occurs in the mitochondria. The primary purpose of photosynthesis is to manufacture the food that plants need to carry out their life processes or store for later use by the plant. The primary purpose of cellular respiration is to release energy from the food plants make to carry out their life processes. The important ideas for this probe are that photosynthesis requires light for the full process to occur and that respiration occurs continuously regardless of whether or not light is present.

Curricular and Instructional Considerations

Elementary Students

In the elementary grades, students learn that plants need water, sunlight, nutrients, and air. They are introduced to the idea that plants make their own food and need oxygen from the air. However, the details of the processes of photosynthesis and respiration should wait until middle school.

Middle School Students

In middle school, students build upon their elementary understanding that plants need sunlight, water, and air to understand the link between these needs and the processes of photosynthesis and respiration. They qualitatively learn that plants take in carbon dioxide and water and use energy from sunlight to produce sugars and oxygen. Details about the process, including formation of ATP and NADPH, are addressed later in high school. They also learn that respiration is a cellular process in which organisms take in oxygen to break down sugars and release energy. However, the way these processes are presented in the curriculum often conveys the misconception that photosynthesis is a plant process and respiration is an animal process, or that the two processes in plants are opposite—that one occurs in the daytime and one occurs at night.

High School Students

High school students build on a basic descriptive understanding of photosynthesis and cellular respiration to understand the cellular and molecular processes involved in synthesizing food and breaking it down to release energy. They move from an understanding of macro structures of a plant that take in and release water and gases and make food to an understanding of the cell organelles involved—chloroplasts and mitochondria. Students may build deeper understanding by examining the specific reactions that occur inside these structures and the arrangement within these structures that facilitates these reactions, including the light and dark reactions of photosynthesis and aerobic and anaerobic cellular respiration.

Administering the Probe

This probe is appropriate for middle and high school students who have prior knowledge of the general processes of photosynthesis and respiration. It is also useful to give to high school students after they have learned about the "light and dark" reactions of photosynthesis to check for misconceptions. See the vignette in the introduction to this book on pages 1–6 for an example of how to use the "sticky bars" FACT with this assessment probe.

Related Ideas in *National Science Education Standards* (NRC 1996)

K–4 The Characteristics of Organisms

- Organisms have basic needs. For example, animals need air, water, and food; plants require air, water, nutrients, and light.

5–8 Structure and Function in Living Systems

- Cells carry out the many functions needed to sustain life.

5–8 Populations and Ecosystems

- Plants and some microorganisms are producers—they make their own food.
- For ecosystems, the major source of energy is sunlight. Energy entering ecosystems as sunlight is transferred by producers into chemical energy by photosynthesis.

9–12 Matter, Energy, and Organization in Living Systems

- Plants capture energy by absorbing light and using it to form strong (covalent) chemical bonds between the atoms of carbon-containing (organic) molecules. These molecules can be used to assemble larger molecules with biological activity (including proteins, DNA, sugars, and fats). In addition, the energy stored in bonds between the atoms (chemical energy) can be used as sources of energy for life processes.
- The chemical bonds of food molecules contain energy. Energy is released when the bonds of food molecules are broken and new compounds with lower energy bonds are formed. Cells usually store this energy temporarily in phosphate bonds of a small, high-energy compound called ATP.

Related Ideas in *Benchmarks for Science Literacy* (AAAS 2009)

K–2 Flow of Matter and Energy

- Plants and animals both need to take in water, and animals need to take in food. In addition, plants need light.

3–5 Flow of Matter and Energy

- Some source of "energy" is needed for all organisms to stay alive and grow.

6–8 Flow of Matter and Energy

- Plants use the energy from light to make sugars from carbon dioxide and water.
- Organisms get energy from oxidizing their food, releasing some of its energy as thermal energy.

Related Research

- Some students think of respiration as synonymous with breathing and therefore do not think of plants as respiring (Driver et al. 1994).
- Several studies reveal that students think of respiration as being the opposite of photosynthesis: with photosynthesis occurring during the day and respiration occurring at night (Driver et al. 1994).

Suggestions for Instruction and Assessment

- Combine this probe with "Respiration" from *Uncovering Student Ideas in Science, Vol. 3: Another 25 Formative Assessment Probes* (Keeley, Eberle, and Dorsey 2008).

- At the high school level, avoid using the term *dark reaction of photosynthesis*. Instead, it is better to use the term *light-independent reaction* so that students don't think photosynthesis can take place in the dark. Sometimes, the multiple terms used confuse students—*light-independent reactions, Calvin cycle,* and *carbon fixation* all mean the same thing. It may be best to use the term *carbon fixation* so that students get the idea that plants need to get carbon from somewhere to make sugar, which is a carbon-based molecule. This reinforces the need to realize the necessity of carbon dioxide as a carbon source.
- Avoid teaching photosynthesis and respiration as opposite processes. Although photosynthesis and respiration are "opposite" in terms of their chemical equations, they are not truly opposite processes in terms of when or where they occur. For example, photosynthesis takes place in the light; respiration in the light and dark. Plant cells have chloroplasts for photosynthesis and mitochondria for cellular respiration; animal cells have mitochondria for cellular respiration but no chloroplasts, because they do not photosynthesize.
- A kinesthetic activity that shows that photosynthesis and respiration are not really opposite processes can be done by handing out cards with the components of the two simplified equations for photosynthesis and respiration and have the students arrange themselves (while standing) to form first one equation or the other. Then have them rearrange themselves to form the other equation. This activity makes it evident which molecules just move from one side of the equation to the other and which parts are different (light energy needed for photosynthesis, ATP produced in respiration, and light-dependent reactions). Collect the cards and do it again with different students holding different components of the equations. Repeat several times, sometimes starting with photosynthesis and sometimes with respiration.

Related NSTA Science Store Publications, NSTA Journal Articles, NSTA SciGuides, NSTA SciPacks, and NSTA Science Objects

American Association for the Advancement of Science (AAAS). 2001. *Atlas of science literacy.* Vol. 1. (See "Flow of Matter in Ecosystems" map, pp. 76–77, and "Flow of Energy in Ecosystems" map, pp. 78–79.) Washington, DC: AAAS.

Koba, S., with A. Tweed. 2009. *Hard-to-teach biology concepts: A framework to deepen student understanding.* (See Chapter 4, "Photosynthesis and Respiration," pp. 119–141.) Arlington, VA: NSTA Press.

Mundry, S., P. Keeley, and C. Landel. 2009. *A leader's guide to science curriculum topic study.* (See Module B6, Photosynthesis and Respiration Facilitation Guide, pp. 144–149.) Thousand Oaks, CA: Corwin Press.

Weinburgh, M. 2004. Teaching photosynthesis: More than a lecture but less than a lab. *Science Scope* 27 (9): 15–17.

Related Curriculum Topic Study Guide (in Keeley 2005)
"Photosynthesis and Respiration"

References

American Association for the Advancement of Science (AAAS). 2009. Benchmarks for science literacy online. *www.project2061.org/publications/bsl/online*

Driver, R., A. Squires, P. Rushworth, and V. Wood-Robinson. 1994. *Making sense of secondary science: Research into children's ideas.* London: RoutledgeFalmer.

Keeley, P. 2005. *Science curriculum topic study: Bridging the gap between standards and practice.* Thousand Oaks, CA: Corwin Press and Arlington, VA: NSTA Press.

Keeley, P., F. Eberle, and C. Dorsey. 2008. *Uncovering student ideas in science, vol. 3: Another 25 formative assessment probes.* Arlington, VA: NSTA Press.

National Research Council (NRC). 1996. *National science education standards.* Washington, DC: National Academies Press.

Food for Corn

Eight farmers were talking about their cornfields. They each had different ideas about the food their corn needed to grow. This is what they said:

Mrs. Farrin: "My corn plants use sunlight as their food."

Mrs. Tobias: "My corn plants use food they get from the soil."

Mr. Cullenberg: "My corn plants use sugar as their food."

Mr. King: "My corn plants use food from the fertilizer I give them."

Mrs. Joslyn: "My corn plants use carbon dioxide and water as their food."

Mr. Cody: "My corn plants use food from the chlorophyll in their leaves."

Mr. Trask: "My corn plants use food from the ears of corn they produce."

Mrs. Ahlholm: "My corn plants don't use food; instead, they make food for animals to eat."

Which farmer do you agree with the most? ________________ Explain why you agree with that farmer.

Food for Corn

Teacher Notes

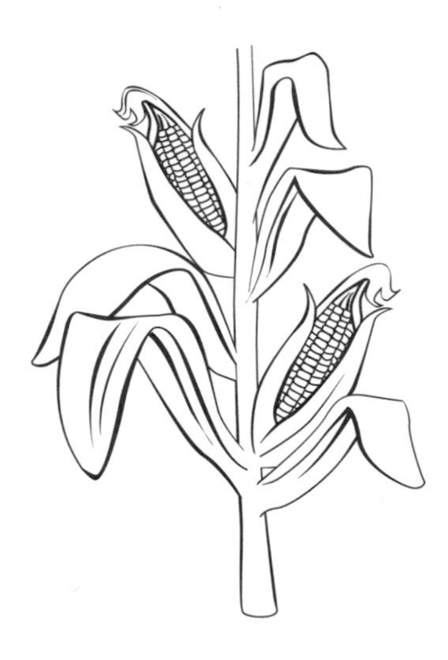

Purpose

The purpose of this assessment probe is to elicit students' ideas about "plant food." The probe is designed to reveal whether students recognize that plants use the sugars they make as their food.

Related Concepts

Food, photosynthesis, plants

Explanation

The best response is Mr. Cullenberg's: "My corn plants use sugar as their food." Plants take in carbon dioxide through their leaves and water through their roots and use energy from sunlight to break apart and reassemble these molecules into a carbohydrate in the form of a simple sugar. This simple sugar (glucose) is often converted to a more complex sugar such as sucrose, starch, or cellulose. Plants can use the food (sugars) they make immediately or store it for later use. Sunlight is the source of energy for the food-making process of photosynthesis, but, by itself, it is not food (Mrs. Farrin's response). The captured energy from sunlight is transformed into chemical energy, which is stored in the chemical bonds between the atoms of carbon-containing organic molecules. Soil is the substrate in which plants grow and from which they take in water and minerals through their roots (Mrs. Tobias's response). Water and minerals are essential nutrients but they are not food. Fertilizer provides essential elements plant need such as potassium, nitrogen, magnesium, and phosphorus, but it is not food (Mr. King's response).

Carbon dioxide and water are substances plants take in to make their food, but, in their original form, they are not food (Mrs. Joslyn's response). They are broken down into atoms and molecules that are reassembled to form the sugar that becomes the food for the plant. Chlorophyll is the green pigment found within the plant cell's chloroplast that absorbs light energy for use during photosynthesis (Mr. Cody's response). Plants convert sugars they make into new structures, such as the ears of

corn that are produced when the corn is pollinated and seeds form. The sugar in the kernels of corn will provide food for the corn seedling as it germinates, but it is not the food the adult corn plant uses to grow (Mr. Trask's response). Plants not only make food but they use it too (Mrs. Ahlholm's response). Food is the material that provides energy and the building blocks for growth and repair for all organisms, including plants.

Curricular and Instructional Considerations

Elementary Students

In the elementary grades, students learn that plants need water, sunlight, nutrients, and air. They may be introduced to the idea that plants make their own food, but the ideas related to the reactants, products, and process of photosynthesis can wait until middle school. Elementary students should know that all living things, including plants, need food and that food provides energy.

Middle School Students

In middle school, students are introduced to the basic process of photosynthesis. They learn that plants make sugars from carbon dioxide that they take in through their leaves and water they take in from their roots and that the process requires energy from sunlight. They learn that plants use the sugars they make as their food in addition to making that food available to other organisms. They learn that food provides a source of energy as well as material for plant growth and repair. They learn that the food that plants make can be used immediately or stored for later use, such as in the bulb of a flowering plant. Middle school is also the time when students develop a scientific conception of *food* that differs from the common, everyday use of the word *food*.

High School Students

In high school, students deepen their understanding of photosynthesis at the molecular level by connecting their growing understanding of chemistry to the biological process of photosynthesis. They learn about the different types of carbohydrates formed by plants from simple sugars. They make connections among sunlight, food, and the energy released from the chemical bonds formed between atoms that make up the molecules of sugar.

Administering the Probe

This probe is best used at the middle and high school level. To activate their thinking, ask students if they have ever seen a corn plant or a field of corn. Where do these corn plants get the food they need to grow? For high school students, you may consider substituting the word *carbohydrates* for *sugar*.

Related Ideas in *National Science Education Standards* (NRC 1996)

K–4 The Characteristics of Organisms

- Organisms have basic needs. For example, animals need air, water, and food; plants require air, water, nutrients, and light.

5–8 Populations and Ecosystems

- Plants and some microorganisms are producers—they make their own food.
- For ecosystems, the major source of energy is sunlight. Energy entering ecosystems as sunlight is transferred by producers into chemical energy by photosynthesis.

9–12 The Cell

★ Plant cells contain chloroplasts, the site of photosynthesis. Plants and many microorganisms use solar energy to combine molecules of carbon dioxide and water into complex, energy-rich organic compounds

★ Indicates a strong match between the ideas elicited by the probe and a national standard's learning goal.

and release oxygen to the environment. This process of photosynthesis provides a vital connection between the Sun and the energy needs of living systems.

9–12 Matter, Energy, and Organization in Living Systems

- The energy for life primarily derives from the Sun. Plants capture energy by absorbing light and using it to form strong (covalent) chemical bonds between the atoms of carbon-containing (organic) molecules. These molecules can be used to assemble larger molecules with biological activity (including proteins, DNA, sugars, and fats). In addition, the energy stored in bonds between the atoms (chemical energy) can be used as sources of energy for life processes.
- The chemical bonds of food molecules contain energy. Energy is released when the bonds of food molecules are broken and new compounds with lower energy bonds are formed. Cells usually store this energy temporarily in phosphate bonds of a small, high-energy compound called ATP.

Related Ideas in *Benchmarks for Science Literacy* (AAAS 2009)

K–2 Flow of Matter and Energy in Ecosystems

- Plants and animals both need to take in water, and animals need to take in food. In addition, plants need light.

K–2 Cells

- Most living things need water, food, and air.

3–5 Flow of Matter and Energy in Ecosystems

- Some source of "energy" is needed for all organisms to stay alive and grow.

6–8 Flow of Matter and Energy in Ecosystems

- Food provides molecules that serve as fuel and building material for all organisms.
- ★ Plants use the energy from light to make sugars from carbon dioxide and water.
- Plants can use the food they make immediately or store it for later use.

9–12 Flow of Matter and Energy in Ecosystems

- The chemical elements that make up the molecules of living things pass through food webs and are combined and recombined in different ways. At each link in a food web, some energy is stored in newly made structures, but much is dissipated into the environment. Continual input of energy from sunlight keeps the process going.

Related Research

- Universally, the most persistent notion people of all ages have about where plants get their food is that plants take their food from the environment, particularly the soil. Students also believe that plants have multiple sources of food (Driver et al. 1994).
- Children appear to consider food as anything useful taken into an organism's body, including water, minerals, and, in the case of plants, carbon dioxide or even sunlight. Typically, students do not consider starch as food for plants (Driver et al. 1994).
- In a study by Tamir (1989), some students thought sunlight, associated with energy, was the food for plants. Many students also considered minerals taken in from the soil as food.

★ Indicates a strong match between the ideas elicited by the probe and a national standard's learning goal.

- Some children consider chlorophyll to be a food substance (Driver et al. 1994). For plants, chlorophyll is not food; it only serves as food for animals that consume it and break it down.
- Food is colloquially understood to be anything an organism takes in for nourishment; therefore, students believe that anything absorbed from the soil is food. This is an instance where everyday meaning and scientific meaning clash and create confusion. Garden fertilizers labeled "plant food" reinforce this erroneous idea that fertilizer is food for plants (Allen 2010).
- The common misconception that plants get their food from the environment rather than manufacturing it internally, and that food for plants is taken in from the outside, is particularly resistant to change, even after instruction (Anderson, Sheldon, and Dubay 1990).
- Some students who know that plants make their own food think they do so for the benefit of animals that eat plants. They do not recognize that plants use the food they make (Driver et al. 1994).

Suggestions for Instruction and Assessment

- Combine this probe with "Is It Food for Plants?" in *Uncovering Student Ideas in Science, Vol. 2: 25 More Formative Assessment Probes* (Keeley, Eberle, and Tugel 2007).
- Combine this probe with probe #9 in this book, "Apple Tree," to determine whether students recognize both the plant structure in which food is made and what plants use their food for.
- Take the time to elicit students' definitions of the word *food;* many students use this word in a way that is not consistent with its biological meaning (AAAS 2009). For example, many students think of food as something that is taken in through the mouth. Have students identify the difference between the everyday use of the word *food* and the scientific use of the word. Contrasting the two uses and providing examples may help students recognize the difference and know when the word is used in a biological context.
- Beginning in elementary grades, students need to recognize that all organisms need and use food. This includes plants as well as animals.
- Understanding that the food plants make is very different from nutrients they take in may be a prerequisite for understanding the idea that plants make their food rather than acquire it from their environment (Roth, Smith, and Anderson 1983).
- High school students can often define *photosynthesis,* provide the equation, name the cell organelle involved, and identify glucose as the food that is made. However, students are rarely asked basic questions that call on them to apply this understanding. Develop questions that use the concept of photosynthesis to explain food, growth, repair, and energy-related plant ideas.
- Provide examples that help students recognize that plants use the sugars they make, such as the maple sap flowing in a tree in the spring or the stored starch in a bulb.
- With high school students, trace disaccharides (sucrose) and polysaccharides (starch, cellulose) in plants back to the simple sugar produced during photosynthesis. Use the analogy of a plant as a kind of food factory, manufacturing its own glucose from raw materials, using it as fuel, converting it to cellulose to build plant structures, or storing it for later use in the form of starch.
- It might be worth pointing out that some common plants do partition part of their sugars to serve as food for animals. Apples, cherries, and some other fleshy fruits evolved specifically to attract animals that

will eat the fruit. The seeds within these fruits can survive passing through the animal's digestive tract and are deposited with feces to grow new plants in new areas. The primary dispersal mechanism of species of plants with fleshy fruits is consumption by animals. Some more astute students might raise this question. Some plants even fine-tune their fruits to ensure only the correct species eat them. The capsaicins that make chili peppers hot are meant to deter mammals (whose guts destroy the seeds) from eating the fruits. Birds cannot sense the chemicals, so they eat them and distribute the seeds.

- The Phenomena and Representations for the Instruction of Science in Middle Schools (PRISMS) website, a collection in the National Science Digital Library (NSDL) funded by the National Science Foundation, provides examples of reviewed web-based phenomena and representations related to the topic of photosynthesis at *http://prisms.mmsa.org*.

Related NSTA Science Store Publications, NSTA Journal Articles, NSTA SciGuides, NSTA SciPacks, and NSTA Science Objects

American Association for the Advancement of Science (AAAS). 2001. *Atlas of science literacy.* Vol. 1. (See "Flow of Matter in Ecosystems" map, pp. 76–77, and "Flow of Energy in Ecosystems" map, pp. 78–79.) Washington, DC: AAAS.

George, R. 2003. How do plants make their own food? *Science and Children* 40 (5): 17.

Koba, S., with A. Tweed. 2009. *Hard-to-teach biology concepts: A framework to deepen student understanding.* (See Chapter 4, "Photosynthesis and Respiration," pp. 119–141.) Arlington, VA: NSTA Press.

Mundry, S., P. Keeley, and C. Landel. 2009. *A leader's guide to science curriculum topic study.* (See Module B6, Photosynthesis and Respiration Facilitation Guide, pp. 144–149.) Thousand Oaks, CA: Corwin Press.

Related Curriculum Topic Study Guides (in Keeley 2005)
"Food and Nutrition"
"Photosynthesis and Respiration"

References

Allen, M. 2010. *Misconceptions in primary science.* Berkshire, England: Open University Press.

American Association for the Advancement of Science (AAAS). 2009. Benchmarks for science literacy online. *www.project2061.org/publications/bsl/online*

Anderson, C., T. Sheldon, and J. Dubay. 1990. The effects of instruction on college non-majors' conceptions of respiration and photosynthesis. *Journal of Research in Science Teaching* 27: 761–776.

Driver, R., A. Squires, P. Rushworth, and V. Wood-Robinson. 1994. *Making sense of secondary science: Research into children's ideas.* London: RoutledgeFalmer.

Keeley, P. 2005. *Science curriculum topic study: Bridging the gap between standards and practice.* Thousand Oaks, CA: Corwin Press and Arlington, VA: NSTA Press.

Keeley, P., F. Eberle, and J. Tugel. 2007. *Uncovering student ideas in science, vol. 2: 25 more formative assessment probes.* Arlington, VA: NSTA Press.

National Research Council (NRC). 1996. *National science education standards.* Washington, DC: National Academies Press.

Roth, K., E. Smith, and C. Anderson. 1983. *Students' conceptions of photosynthesis and food for plants.* East Lansing, MI: Michigan State University, Institute for Research on Teaching.

Tamir, P. 1989. Some issues related to justifications to multiple choice answers. *Journal of Biological Education* 11 (1): 48–56.

Pumpkin Seeds

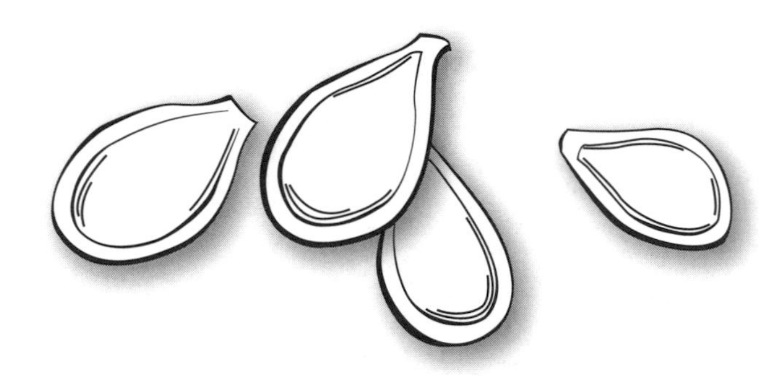

Four friends were planting pumpkin seeds for a class project. They each had different ideas about how the seeds should be placed in the soil. This is what they said:

Nancy: "I think more seeds will sprout if we plant them with the pointed tip down."

Lynn: "I think more seeds will sprout if we plant them with the pointed tip up."

Joyce: "I think more seeds will sprout if we plant them on their flat sides."

Lisa: "I think the way we put the seeds into the soil doesn't matter."

Which friend do you agree with the most? ____________ Explain why you agree.

____

____

____

____

____

____

____

____

____

Pumpkin Seeds

Teacher Notes

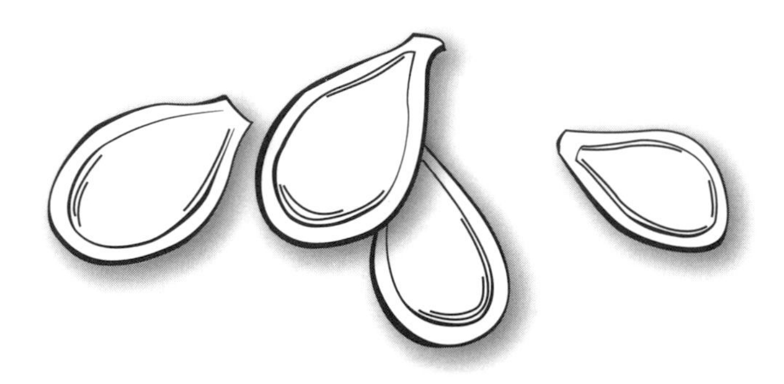

Purpose

The purpose of this assessment probe is to elicit students' ideas about seed germination. The probe is designed to find out whether students recognize that germination will occur under the right conditions regardless of the orientation of the seed.

Related Concepts

Germination, tropism, behavior of organisms, plants, respiration

Explanation

Lisa has the best answer—"I think the way we put the seeds into the soil doesn't matter." Plants have special responses to stimuli, called tropisms. Geotropism (or gravitropism) is a response to direction in relation to the Earth. Regardless of the orientation of the seed, roots grow in the direction of the gravitational force—a downward direction toward the Earth's center. This is called a positive geotropism. The upper part of the seedling, the shoot, exhibits a negative geotropism. Regardless of the seed's orientation, the shoot will grow away from the Earth, in an upward direction. Interestingly, Charles Darwin and his son Francis were among the first scientists to document positive geotropism of roots and negative geotropism of shoots when they published *The Power of Movement in Plants* in 1880.

Curricular and Instructional Considerations

Elementary Students

Planting seeds to learn about the needs of a seed and a growing plant are common inquiry-based activities in the elementary grades. Students learn about the physical characteristics of plants and animals and the observable behaviors that allow them to live in their environments. The behaviors of plants are not as obvious to students as are the behaviors of animals; therefore, they need experiences with both types of organisms. Students learn how to

test their ideas to investigate phenomena such as whether planting seeds in different positions would make a difference in their germination.

Middle School Students

In middle school, students begin to learn about the behavior of plants and animals in response to internal and external stimuli. At this level they also recognize that "response to stimuli" is one of the criteria for determining whether an object is living or nonliving (although they may recognize this more readily with animals than with plants). They learn about various plant tropisms as movement responses by plants to certain environmental stimuli. At this level they can connect their understanding of Earth's gravity with the gravitropism of plants to understand why it does not matter which way the seed is planted.

High School

At this level students are more systematic in investigating plant responses to stimuli such as gravity. Their increased knowledge of plant physiology allows them to investigate how and where growth occurs in roots as well as the important role played by plant hormones (auxins) in a variety of tropisms. They differentiate between the negative gravitropism displayed by the shoot and the positive gravitropism displayed by the root.

Administering the Probe

This probe is best used at the elementary level up through early middle school. If pumpkin seeds are available, show students the seeds so they can see they are flat and have a pointed tip.

Related Ideas in *National Science Education Standards* (NRC 1996)

K–4 The Characteristics of Organisms

- The behavior of individual organisms is influenced by internal cues (such as hunger) and by external cues (such as a change in the environment).

5–8 Regulation and Behavior

★ Behavior is one kind of response an organism can make to an internal or environmental stimulus.

9–12 The Behavior of Organisms

★ Organisms have behavioral responses to internal changes and to external stimuli. Responses to external stimuli can result from interactions with the organism's own species and others, as well as environmental changes; these responses either can be innate or learned. The broad patterns of behavior exhibited by animals have evolved to ensure reproductive success. Animals often live in unpredictable environments, and so their behavior must be flexible enough to deal with uncertainty and change. Plants also respond to stimuli.

Related Ideas in *Benchmarks for Science Literacy* (AAAS 2009)

K–2 Flow of Matter and Energy

- Plants and animals both need to take in water, and animals need to take in food. In addition, plants need light.

Related Research

- There is scant research on children's ideas related to plant tropisms. However, there have been studies that show children use movement and response as criteria to decide whether something is living. As a result, some children do not think plants are living things because they do not recognize their movement and behavioral responses to the environment (Driver et al. 1994).

★ Indicates a strong match between the ideas elicited by the probe and a national standard's learning goal.

Suggestions for Instruction and Assessment

- This probe can be combined with "Plants in the Dark and Light" from *Uncovering Student Ideas in Science, Vol. 2: 25 More Formative Assessment Probes* to elicit students' ideas about phototropism (Keeley, Eberle, and Tugel 2007).
- This probe can be used as a P-E-O probe (Keeley 2008). Have students make *predictions* and *explain* the reason for their predictions. Then provide the students with seeds to test their predictions by making *observations*. When students find their observations do not match their predictions, encourage them to reconsider their explanations.
- Provide students with different types of seeds, besides the pumpkin seeds, to test their ideas.
- Plant tropisms are a fundamental concept in plant science and can be used as evidence of observable signs of life in plants.
- Use caution when using the term *geotropism*. It may be misleading to students because the stimulus is gravity, not the Earth. A better term is *gravitropism*.
- Be aware that many students think plants are simple and "primitive" and do very little. Opportunities to study plant hormones, other chemical signals, and tropisms help students see that plants are highly sophisticated and responsive to their environments. Plants rival animals in their ability to respond and adapt to an uncertain environment. Point out to students that they undergo physiological responses to drought, temperature changes, pathogens, and injuries; shift their growth pattern or undergo tropisms and nastic movements in response to water levels, nutrient availability, and environmental cues; exhibit photoperiodicity and circadian rhythms; react biochemically to the presence of other species or members of their own species; and compete for limited resources.

Related NSTA Science Store Publications, NSTA Journal Articles, NSTA SciGuides, NSTA SciPacks, and NSTA Science Objects

Konicek-Moran, R. 2010. Halloween science. In *Even more everyday science mysteries: Stories for inquiry-based science teaching*, 125–134. Arlington, VA: NSTA Press.

Tolman, M., and G. Hardy. 1999. Teaching tropisms. *Science and Children* 37 (3): 14–17.

Related Curriculum Topic Study Guide (in Keeley 2005)
"Behavioral Characteristics of Organisms"

References

American Association for the Advancement of Science (AAAS). 2009. Benchmarks for science literacy online. *www.project2061.org/publications/bsl/online*

Driver, R., A. Squires, P. Rushworth, and V. Wood-Robinson. 1994. *Making sense of secondary science: Research into children's ideas.* London: RoutledgeFalmer.

Keeley, P. 2005. *Science curriculum topic study: Bridging the gap between standards and practice.* Thousand Oaks, CA: Corwin Press and Arlington, VA: NSTA Press.

Keeley, P. 2008. *Science formative assessment: 75 practical strategies for linking assessment, instruction, and learning.* Thousand Oaks, CA: Corwin Press and Atlington, VA: NSTA Press.

Keeley, P., F. Eberle, and J. Tugel. 2007. *Uncovering student ideas in science, vol. 2: 25 more formative assessment probes.* Arlington, VA: NSTA Press.

National Research Council (NRC). 1996. *National science education standards.* Washington, DC: National Academies Press.

Rocky Soil

Natalie and her brothers were planting a vegetable garden in their backyard. They noticed there were a lot of rocks in the soil. They wondered whether they should dig out all the rocks before planting their seeds. This is what they said:

Eric: "I think we can leave most of the rocks in. The roots will grow right through the rocks."

Ethan: "I think we can leave most of the rocks in. The roots will just grow right around the rocks."

Natalie: "I think we need to take most of the rocks out or else they will stop the roots from growing down into the soil."

Which friend do you agree with the most? ____________ Explain why you agree.

____

____

____

____

____

____

____

____

____

Rocky Soil

Teacher Notes

Purpose

The purpose of this assessment probe is to elicit students' ideas about plant behavior. The probe is designed to find out whether students recognize that the roots of a plant will grow around an obstructing object.

Related Concepts

Plants, tropism, behavior of organisms

Explanation

Ethan has the best answer: "I think we can leave most of the rocks in. The roots will just grow right around the rocks." Plants have directional behavioral responses to environmental stimuli, called tropisms. Thigmotropism is a response to touch or contact with objects. Geotropism is a response to gravity, which is why plant roots grow downward. Roots depend on these tropisms as they grow downward through the soil. When root tips touch obstructing objects like stones and rocks in their path, they change their direction of growth to pass around the object (but not growing upward). This is called negative thigmotropism, because the net result is that the root tip grows away from the rock that touches it. Some plants, like peas, beans, and vines, will respond to touch in a positive way, by moving toward the object. For example, peas have tendrils that coil around an object when they touch it. This helps provide support for the growing plant.

Curricular and Instructional Considerations

Elementary Students

Planting seeds and growing plants are common inquiry-based activities in the elementary grades, leading to an understanding of the needs of organisms, characteristics of organisms, and life cycles. Students learn about the physical characteristics of plants and animals and the observable behaviors that allow them to live in their environments. They also learn how to test their ideas

to investigate observable phenomena such as whether placing objects in the path of a root would make a difference in the growth or direction of growth of the root.

Middle School Students

In middle school, students begin to learn about the behavior of plants and animals in response to internal and external stimuli. They also recognize "response to stimuli" as one of the criteria for determining whether an object is living or nonliving (although they may recognize this criterion more readily with animals than with plants.) They learn about various plant tropisms, such as thigmotropism, as ways that plants respond to their environments.

High School Students

At this level students are more systematic in investigating plant responses to physical stimuli. They can draw on their increased knowledge of plant physiology to investigate how plants respond to touch and explore the interaction between gravitropism and thigmotropism.

Administering the Probe

This probe is most appropriate to use with K–8 students, especially in an inquiry context in which they have opportunities to investigate plant behavior. It is also appropriate to use at the high school level, where students are deepening their understanding of plant behaviors such as their response to stimuli.

Related Ideas in *National Science Education Standards* (NRC 1996)

K–4 The Characteristics of Organisms

- The behavior of individual organisms is influenced by internal cues (such as hunger) and by external cues (such as a change in the environment).

5–8 Regulation and Behavior

★ Behavior is one kind of response an organism can make to an internal or environmental stimulus.

9–12 The Behavior of Organisms

★ Organisms have behavioral responses to internal changes and to external stimuli. Responses to external stimuli can result from interactions with the organism's own species and others, as well as environmental changes; these responses either can be innate or learned. The broad patterns of behavior exhibited by animals have evolved to ensure reproductive success. Animals often live in unpredictable environments, and so their behavior must be flexible enough to deal with uncertainty and change. Plants also respond to stimuli.

Related Ideas in *Benchmarks for Science Literacy* (AAAS 2009)

K–2 Flow of Matter and Energy

- Plants and animals both need to take in water, and animals need to take in food. In addition, plants need light.

Related Research

- There is scant research on children's ideas related to plant tropisms. However, there have been studies that show children use movement and response as criteria to decide whether something is living. As a result, some children do not think plants are living things because they do not recognize their movement and behavioral responses to the environment (Driver et al. 1994).

Suggestions for Instruction and Assessment

- This probe can be combined with "Plants in the Dark and Light" from *Uncovering*

★ Indicates a strong match between the ideas elicited by the probe and a national standard's learning goal.

Student Ideas in Science, Vol. 2: 25 More Formative Assessment Probes to elicit students' ideas about a different tropism—phototropism (Keeley, Eberle, and Tugel 2007).

- This probe can be used as a P-E-O probe (Keeley 2008). Have students make *predictions* and *explain* the reasons for their predictions. Then provide the students with seeds to test their predictions by making *observations*. The students can plant the seeds with obstructions right below where the seeds were planted. Later, after seedlings have grown, students can remove the soil and observe how the roots grew away from the obstruction. When students find their observations do not match their predictions, encourage them to reconsider their explanation.
- Be aware that many students think plants are simple and "primitive" and do very little. Opportunities to study plant hormones, other chemical signals, and tropisms help students see that plants are highly sophisticated and responsive to their environments. Plants rival animals in their ability to respond and adapt to uncertain environments. Point out to students that plants undergo physiological responses to drought, temperature changes, pathogens, and injuries; shift their growth pattern or undergo tropisms and nastic movements in response to water levels, nutrient availability, and environmental cues; exhibit photoperiodicity and circadian rhythms; react biochemically to the presence of other species or members of their own species; and compete for limited resources.
- The National Gardening Association has a website with activities for exploring plant tropisms. Students can grow "crazy carrots" to observe the response of roots to objects in their path (*www.kidsgardening.com/growingideas/projects/july04/pg2.html*).

Related NSTA Science Store Publications, NSTA Journal Articles, NSTA SciGuides, NSTA SciPacks, and NSTA Science Objects

Konicek-Moran, R. Forthcoming. The new greenhouse. In *Yet more everyday science mysteries: Stories for inquiry-based science teaching.* Arlington, VA: NSTA Press.

Tolman, M., and G. Hardy. 1999. Teaching tropisms. *Science and Children* 37 (3): 14–17.

Related Curriculum Topic Study Guide (in Keeley 2005)
"Behavioral Characteristics of Organisms"

References

American Association for the Advancement of Science (AAAS). 2009. Benchmarks for science literacy online. *www.project2061.org/publications/bsl/online*

Driver, R., A. Squires, P. Rushworth, and V. Wood-Robinson. 1994. *Making sense of secondary science: Research into children's ideas.* London: RoutledgeFalmer.

Keeley, P. 2005. *Science curriculum topic study: Bridging the gap between standards and practice.* Thousand Oaks, CA: Corwin Press.

Keeley, P. 2008. *Science formative assessment: 75 practical strategies for linking assessment, instruction, and learning.* Thousand Oaks, CA: Corwin Press and Arlington, VA: NSTA Press.

Keeley, P., F. Eberle, and J. Tugel. 2007. *Uncovering student ideas in science, vol. 2: 25 more formative assessment probes.* Arlington, VA: NSTA Press.

National Research Council (NRC). 1996. *National science education standards.* Washington, DC: National Academies Press.

Section 2

Ecosystems and Adaptation; Reproduction, Life Cycles, and Heredity; Human Biology

Concept Matrix:
Ecosystems and Adaptation; Reproduction, Life Cycles, and Heredity; Human Biology
Probes #14–#25

PROBES	14. Is It a Consumer?	15. Food Chain Energy	16. Ecosystem Cycles	17. No More Plants	18. Changing Environment	19. Eggs	20. Chrysalis	21. DNA, Genes, and Chromosomes	22. Eye Color	23. Human Body	24. Human Excretory System	25. Antibiotics
GRADE-LEVEL USE →	3–12	6–12	6–12	K–8	3–12	3–8	K–5	6–12	9–12	3–8	6–12	6–12
RELATED CONCEPTS ↓												
adaptation					X							
animals	X											
antibiotics												X
cells										X		
chromosome								X				
chrysalis							X					
consumer	X	X		X								
cycling of matter			X									
digestive system											X	
DNA								X				
ecosystem			X	X								
egg						X						
excretion											X	
excretory system											X	
flow of energy		X	X									
food		X										
food chain	X	X		X								
food web	X	X		X								
genes								X	X			
human body										X		
infectious disease												X
interdependence				X	X							
life cycle						X	X					
living							X					
metabolic waste											X	
natural selection					X							
producer	X	X		X								
pupa							X					
secretion											X	
sexual reproduction						X						
traits									X			
virus												X

Is It a Consumer?

Some living things are called *consumers.* Check off each organism on the list that can be considered a consumer.

_____ human	_____ butterfly	_____ squirrel
_____ lion	_____ cow	_____ hawk
_____ spider	_____ tree	_____ pigeon
_____ grass	_____ fly	_____ fox
_____ dog	_____ ant	_____ rabbit
_____ shark	_____ rosebush	_____ snail

Explain your thinking. What rule or reasoning did you use to decide if an organism is a consumer?

____

____

____

____

____

____

Is It a Consumer?

Teacher Notes

Purpose

The purpose of this assessment probe is to elicit students' ideas about consumers. The probe is designed to see how students interpret the word *consumer* in a biological context.

Related Concepts

Consumer, producer, food chain, food web, animals

Explanation

All of the organisms on the list are consumers except for the tree, grass, and rosebush. By definition, a consumer is an organism that must eat other organisms to obtain food. Another name for a consumer organism is *heterotroph*. In contrast, producers are organisms that make sugars through photosynthesis. The sugars are used to produce organic compounds, which in turn are necessary to the producers' growth and metabolism. Producers include plants and photosynthetic microbes. Another name for a producer organism is *autotroph*. Consumers eat producers or other consumers and can be herbivores, carnivores, or omnivores. Decomposers (such as some fungi and bacteria) consume other organisms that have died.

Curricular and Instructional Considerations

Elementary Students

In the elementary grades, students are introduced to simple food chains. They learn who eats what and distinguish between plants, which make their own food, and animals, which acquire their food by eating other animals or plants. They observe interactions among organisms in their immediate environments.

Middle School Students

In middle school, students move from food chains to food webs, understanding that a food chain is a part of a food web. They are introduced to the terminology that describes the flow of energy in ecosystems from producers

to consumers to decomposers. Their understanding of food relationships among organisms is applied to their understanding of ecosystems as they trace the cycling of matter through ecosystems.

High School Students

High school students develop a deeper understanding of ecosystems and the complex interaction of its parts, including matter cycling and energy flow. Their understanding of food relationships in ecosystems is connected to their growing idea of systems. By high school, students should be familiar with terms such as *producer*, *consumer*, and *decomposer* to describe matter and energy relationships in ecosystems.

Administering the Probe

This probe can be used with elementary, middle, and high school students who have previously encountered the word *consumer* in a life science context. Make sure students are familiar with the organisms on the list on page 85. Remove any organisms that students are not familiar with or provide pictures that show the organism. This probe can be administered as a card sort (Keeley 2008). Write the name of each organism on a card. Give the set of cards (you might want to make three or four sets) to a student group. The students sort the cards into two piles—one made up of organisms they think are consumers and the other of organisms they think are not consumers. As they sort their cards, they must provide justification for their groupings.

Related Ideas in *National Science Education Standards* (NRC 1996)

K–4 Organisms and Their Environments

- All animals depend on plants. Some animals eat plants for food. Other animals eat animals that eat the plants.

5–8 Populations and Ecosystems

★ Populations of organisms can be categorized by the function they serve in an ecosystem. Plants and some microorganisms are producers—they make their own food. All animals, including humans, are consumers, which obtain food by eating other organisms. Decomposers, primarily bacteria and fungi, are consumers that use waste materials and dead organisms for food. Food webs identify the relationships among producers, consumers, and decomposers in an ecosystem.

9–12 The Interdependence of Organisms

- Energy flows through ecosystems in one direction, from photosynthetic organisms to herbivores to carnivores and decomposers.

Related Ideas in *Benchmarks for Science Literacy* (AAAS 2009)

K–2 Flow of Matter and Energy

- Plants and animals both need to take in water, and animals need to take in food. In addition, plants need light.

K–2 Interdependence of Life

- Animals eat plants or other animals for food and may also use plants (or even other animals) for shelter and nesting.

3–5 Flow of Matter and Energy

- Almost all kinds of animals' food can be traced back to plants.

6–8 Flow of Matter and Energy

- Plants can use the food they make immediately or store it for later use.

★ Organisms that eat plants break down the plant structures to produce the materials

★ Indicates a strong match between the ideas elicited by the probe and a national standard's learning goal.

and energy they need to survive. Then they are consumed by other organisms.
- Almost all food energy originally comes from sunlight.

6–8 Interdependence of Life
- Interactions between organisms may be for nourishment, reproduction, or protection and may benefit one of the organisms or both of them.

9–12 Flow of Matter and Energy
- The chemical elements that make up the molecules of living things pass through food webs and are combined and recombined in different ways. At each link in a food web, some energy is stored in newly made structures, but much is dissipated into the environment. Continual input of energy from sunlight keeps the process going.

Related Research

- Bell and Barker's study (1982) found that students had a limited understanding of the terms *producer* and *consumer.* Their recognition of these terms was tied to their understanding of plants and animals. Once the scientific meanings of the words *plant* and *animal* were taught and understood, students could use the terms *producer* and *consumer* appropriately (Driver et al. 1994).

Suggestions for Instruction and Assessment

- As the research suggests, consider developing the biological concepts of plant and animal first and then connecting the terms *producer* and *consumer* to them. Plants make (produce) their own food, which other organisms can eat, which makes them producers. Animals must consume food from plant or animal sources, which makes them consumers. With older students, you can expand this concept to include consumer fungi, bacteria, and protists and producer bacteria and protists.
- Add *consumer* and *producer* to students' growing list of words that show how words used in an everyday context differ from their scientific meanings.
- Consider developing a similar probe, "Is It a Producer?," to elicit students' ideas about producers in an ecological context.
- Once students have a conceptual understanding of what a consumer is, have them re-sort their consumer cards (used in the card sort described on p. 87) into primary, secondary, or tertiary consumers, tracing back the energy to producers each time.

Related NSTA Science Store Publications, NSTA Journal Articles, NSTA SciGuides, NSTA SciPacks, and NSTA Science Objects

American Association for the Advancement of Science (AAAS). 2001. *Atlas of science literacy.* Vol. 1. (See "Flow of Matter in Ecosystems" map, pp. 76–77.) Washington, DC: AAAS.

American Association for the Advancement of Science (AAAS). 2001. *Atlas of science literacy.* Vol. 1. (See "Flow of Energy in Ecosystems" map, pp. 78–79.) Washington, DC: AAAS.

Koba, S., with A. Tweed. 2009. *Hard-to-teach biology concepts: A framework to deepen student understanding.* (See Chapter 4, "Photosynthesis and Respiration," pp. 119–141.) Arlington, VA: NSTA Press.

Konicek-Moran, R. Forthcoming. What did the owl eat? In *Yet more everyday science mysteries: Stories for inquiry-based science teaching.* Arlington, VA: NSTA Press.

Staires, J. 2007. Science sampler: Word wall connections. *Science Scope* 30 (5): 64–65.

Warren, P., and J. Galle. 2005. *Exploring ecology: 49 ready-to-use activities for grades 4–9.* Arlington, VA: NSTA Press.

Related Curriculum Topic Study Guides (in Keeley 2005)
"Ecosystems"
"Food Chains and Food Webs"

References

American Association for the Advancement of Science (AAAS). 2009. Benchmarks for science literacy online. *www.project2061.org/publications/bsl/online*

Bell, B., and M. Barker. 1982. Towards a scientific concept of animal. *Journal of Biological Education* 16: 197–200.

Driver, R., A. Squires, P. Rushworth, and V. Wood-Robinson. 1994. *Making sense of secondary science: Research into children's ideas.* London: RoutledgeFalmer.

Keeley, P. 2005. *Science curriculum topic study: Bridging the gap between standards and practice.* Thousand Oaks, CA: Corwin Press and Arlington, VA: NSTA Press.

Keeley, P. 2008. *Science formative assessment: 75 practical strategies for linking assessment, instruction, and learning.* Thousand Oaks, CA: Corwin Press.

National Research Council (NRC). 1996. *National science education standards.* Washington, DC: National Academies Press.

Food Chain Energy

Four friends were arguing about the flow of energy between organisms in a food chain. They wondered what happened to the stored energy in the grass plant as one organism eats another in a food chain relationship. This is what they said:

Tatyana: "I think most of the plant's energy ended up in the fox."

Molly: "I think most of the plant's energy ended up in the snake."

Amos: "I think most of the plant's energy ended up in the grasshopper."

Ursula: "I think the amount of energy that was in the plant stayed the same when it passed through the organism in the food chain."

Which student do you agree with the most? ______________ Explain why you agree.

____

____

____

____

____

____

____

____

____

____

Food Chain Energy

Teacher Notes

Purpose

The purpose of this assessment probe is to elicit students' ideas about the flow of energy in a trophic relationship that children are most familiar with—food chains. The probe is designed to find out if students recognize that most of the energy is concentrated at the beginning of a food chain (i.e., in the producers) and that some energy is lost to the environment each time it is transferred from one organism to another.

Related Concepts

Food, flow of energy, food chain, food web, consumer, producer

Explanation

The best answer is Amos's: "I think most of the plant's energy ended up in the grasshopper."

As one organism is consumed by another, the total amount of energy from that food passed from one level to the next is generally only about 10% of the energy from the previous organism. About 90% of the energy is lost through activities such as movement and other life processes. Much of the lost energy is given off as heat and undigested food that is not passed on. Some of the energy is used for growth or stored for later use and remains in organic matter in the consumer's body. This energy is then transferred to the next consumer when the organism is eaten. Therefore, as you move up a food chain to the predators, there is less of that original plant energy available. Only a tiny fraction of the energy traced back to plants is stored in the predators' tissues. Conversely, at the beginning of the food chain, where the producers are found, there is much more energy available.

This probe deals only with the transfer of energy that came from the single plant organism that was consumed. It does not deal with the total amount of energy in each trophic level. Organisms at higher trophic levels need to eat about 10 times more than organisms at a lower trophic level to get the same amount of

usable food energy. That is why they are represented as a pyramid, with few predators at the top.

Curricular and Instructional Considerations

Elementary Students

In the elementary grades, students are introduced to food chains and learn that all animals' food can be traced back to plants. They can follow a food chain—from plants, to animals that eat plants, to animals that eat other animals, to the decomposers that consume dead organisms. Understanding energy transfer and what happens to the total amount of matter and energy can wait until middle school.

Middle School Students

Most of the emphasis in the middle grades is on following matter as it is transferred through food webs. A food chain represents just one pathway within a larger a food web. By the end of eighth grade, students should begin to link energy flow ideas to their knowledge of the cycling of matter in ecosystems. They should be able to follow the transfer of energy as one organism eats another, explain how animals and plants obtain and use energy, and understand the biological meaning of food. Formal terminology related to food webs is introduced in context at this level.

High School Students

High school students have sufficient knowledge of matter and energy to link these conversions to what happens in living systems. Students account for the energy in the tissues of living organisms as being stored in molecular configurations traced back to photosynthesis and released during oxidation. Although there is no need to account for all the energy, students should know that most of the energy transferred from one organism to another is lost to the environment as heat. They should have opportunities to observe heat generated by consumers and decomposers. They can use mathematics to calculate the percentage of energy passed on from one organism to another. As a result, they recognize there is a limit to the number of organisms that can transfer energy in a food chain and a limit to the total number of organisms in each level of a food pyramid.

Administering the Probe

This probe is most appropriate for middle and high school students because of the energy focus. Make sure students know the probe is referring to the transfer of energy, not the transfer of matter. Also, if you choose to substitute a food chain or food web diagram, make sure students know that the arrow in a food chain or food web refers to the matter and energy in one organism *being transferred* to the next. Some students are confused by the arrows and interpret the arrow as meaning the organism eats the organism it points to. You may substitute the food chain illustration in this probe with one students are more familiar with and adjust the answer choices accordingly. Also, be sure they know that the question is not about which organism has more energy, as there is no question that an individual fox has more energy than a single grass plant or grasshopper. The question is about what happens to the energy from that single plant as its energy is transferred through the food chain.

Related Ideas in *National Science Education Standards* (NRC 1996)

K–4 Organisms and Their Environments

- All animals depend on plants. Some animals eat plants for food. Other animals eat animals that eat the plants.

5–8 Populations and Ecosystems

- For ecosystems, the major source of energy is sunlight. Energy entering ecosystems as sunlight is transferred by producers into chemical energy through photosynthesis. That energy then passes from organism to organism in food webs.

9–12 The Interdependence of Organisms

- Energy flows through ecosystems in one direction, from photosynthetic organisms to herbivores to carnivores and decomposers.

9–12 Matter, Energy, and Organization in Living Systems

- The complexity and organization of organisms accommodates the need for obtaining, transforming, transporting, releasing, and eliminating the matter and energy used to sustain the organism.
- ★ As matter and energy flow through different levels of organization of living systems—cells, organs, organisms, communities—and between living systems and the physical environment, chemical elements are recombined in different ways. Each recombination results in storage and dissipation of energy into the environment as heat. Matter and energy are conserved in each change.

Related Ideas in *Benchmarks for Science Literacy* (AAAS 2009)

K–2 Interdependence of Life

- Animals eat plants or other animals for food and may also use plants (or even other animals) for shelter and nesting.

3–5 Flow of Matter and Energy

- Almost all kinds of animals' food can be traced back to plants.
- Some source of "energy" is needed for all organisms to stay alive and grow.

6–8 Flow of Matter and Energy

- Energy can change from one form to another in living things.
- ★ Organisms get energy from oxidizing their food, releasing some of its energy as thermal energy.
- Almost all food energy originally comes from sunlight.

9–12 Flow of Matter and Energy

- ★ The chemical elements that make up the molecules of living things pass through food webs and are combined and recombined in different ways. At each link in a food web, some energy is stored in newly made structures, but much is dissipated into the environment. Continual input of energy from sunlight keeps the process going.

Related Research

- Some students consider "stronger" organisms as having more energy, which they use to feed on weaker organisms with less energy (Driver et al. 1994).
- Some students use accumulation reasoning to explain what happens to the energy as it flows through a food chain. They believe that energy adds up through an ecosystem, so that a top predator would have all the energy from the other organisms and producers that came before it in the chain (Driver et al. 1994).

Suggestions for Instruction and Assessment

- Food chain diagrams are oversimplistic representations of matter and energy flow and do not show what happens to energy as it flows from one organism to the next. Use food pyramids to visually represent what happens to the energy, with plants at

★ Indicates a strong match between the ideas elicited by the probe and a national standard's learning goal.

the base and the top predator at the top. The food pyramid visually represents the biomass that is needed to be transferred from one level to the next. The shape also represents the dwindling amount of energy available at each level.

- Although this probe uses a simple food chain, make sure that students do not think mass–energy transfer is linear (a chain) rather than a network (web). In middle school, students need to recognize the fine distinction between a part of a web (i.e., a chain) and the web as a whole.
- After administering this probe with the food chain diagram, consider giving it again with a trophic pyramid diagram using the same organisms. Analyze the responses to see if students' ideas changed when presented with the pyramid illustration.
- Visually illustrate the flow of energy from one organism to another using 1 L of yellow-colored water. Have the students imagine that the yellow water contains the energy in a producer such as a plant (but be sure they know this is a model and that energy is not physical matter, such as the colored water). This energy is transferred to a primary consumer when the plant is eaten. Pour out one-tenth of the water (100 ml) into a clear cup and hold it up. This water represents the energy the primary consumer transfers when it is eaten. Now pour one-tenth from the 100 ml cup into a 10 ml cup. This smaller amount of water represents the energy transferred to a secondary consumer when it eats the primary consumer.

  At this point, either a tertiary consumer comes along and eats the secondary consumer or the secondary consumer dies and is consumed by decomposers. Again, transfer one-tenth of the water (1 ml) representing the energy from the secondary consumer. Students will barely be able to see the yellow liquid representing the total energy transferred to the top predator or decomposers. Show them all the remaining liquid that was not transferred that represents the energy lost to the ecosystem, mostly as heat.
- If available energy decreases from one trophic level to the next (as well as biomass), ask students if there is a limit to the number of trophic levels that can exist in an ecosystem. This question can lead to an interesting discussion of what it takes to sustain organisms in an ecosystem.
- PRISMS (*prisms.mmsa.org*) has a collection of reviewed web-based resources that provide examples of phenomena and representations of the flow of energy in ecosystems.

Related NSTA Science Store Publications, NSTA Journal Articles, NSTA SciGuides, NSTA SciPacks, and NSTA Science Objects

American Association for the Advancement of Science (AAAS). 2001. *Atlas of science literacy.* Vol. 1. (See "Flow of Energy in Ecosystems" map, pp. 78–79.) Washington, DC: AAAS.

Konicek-Moran, R. Forthcoming. What did the owl eat? In *Yet more everyday science mysteries: Stories for inquiry-based science teaching.* Arlington, VA: NSTA Press.

Staires, J. 2007. Science sampler: Word wall connections. *Science Scope* 30 (5): 64–65.

Warren, P., and J. Galle. 2005. *Exploring ecology: 49 ready-to-use activities for grades 4–9.* Arlington, VA: NSTA Press.

Related Curriculum Topic Study Guides (in Keeley 2005)
"Food Chains and Food Webs"
"Flow of Energy Through Ecosystems"

References

American Association for the Advancement of Science (AAAS). 2009. Benchmarks for science literacy online. *www.project2061.org/publications/bsl/online*

Driver, R., A. Squires, P. Rushworth, and V. Wood-Robinson. 1994. *Making sense of secondary science: Research into children's ideas*. London: RoutledgeFalmer.

Keeley, P. 2005. *Science curriculum topic study: Bridging the gap between standards and practice.* Thousand Oaks, CA: Corwin Press and Arlington, VA: NSTA Press.

National Research Council (NRC). 1996. *National science education standards*. Washington, DC: National Academies Press.

Ecosystem Cycles

Four friends were talking about how matter and energy move through an ecosystem. This is what they said:

Morrie: "I think only energy cycles through an ecosystem."

Felicia: "I think only matter cycles through an ecosystem."

Stefano: "I think both matter and energy cycle through an ecosystem."

Lincoln: "I think neither matter nor energy cycles through an ecosystem."

Which friend do you most agree with? ______________________ Explain your thinking.

____

____

____

____

____

____

____

____

____

____

____

Ecosystem Cycles

Teacher Notes

Purpose

The purpose of this assessment probe is to elicit students' ideas about the transfer of matter and energy in ecosystems. The probe is designed to reveal whether students recognize that only matter is cycled through an ecosystem.

Related Concepts

Ecosystem, cycling of matter, flow of energy

Explanation

Felicia has the best answer: "I think only matter cycles through an ecosystem." Matter and energy both move through an ecosystem. However, only matter cycles back and forth between organisms and the environment; energy moves only in one direction, with much of it being dissipated into the environment as heat. Both matter and energy can be transferred from one organism to another or from an organism to the environment; but only matter cycles within ecosystems, being used in various forms as it moves through food webs, water, soil, and the atmosphere.

The cycles in which elements or molecules move through living and nonliving components in an ecosystem are referred to as biogeochemical cycles. Examples of common elements or molecules that go through biogeochemical cycles include water, carbon, oxygen, nitrogen, phosphorus, and sulfur. The important thing to remember is that matter *cycles,* energy *flows.* Thus, an ecosystem has a constant need for energy input via photosynthetic organisms capturing the Sun's energy.

Curricular and Instructional Considerations

Elementary Students

In the elementary grades, students learn that some materials, including once-living organisms, are recycled by the earth. They know that ecosystems include living and nonliving matter and that all living things require a source

of energy. They learn that the Sun is the major source of energy in most ecosystems on Earth and that all food can be traced back to plants. They can follow a food chain, from plants, to animals that eat plants, to animals that eat other animals, and to the decomposers that consume dead organisms and recycle once-living material. Understanding matter cycling at a molecular level and energy flow can wait until middle school. The water cycle is the biogeochemical cycle students learn about in elementary grades.

Middle School Students

Most of the emphasis in the middle grades is on following matter as it is transferred through food webs and into nonliving components of ecosystems. By the end of eighth grade, students should begin to link energy flow ideas to their knowledge of the cycling of matter in ecosystems. They should be able to follow the unidirectional transfer of energy as one organism eats another and know that energy does not cycle back into the ecosystem. They build on their previous knowledge of the water cycle by learning about other biogeochemical cycles, including the carbon, oxygen, and nitrogen cycles. Middle school students should also know that matter and energy are not created or destroyed as they move through an ecosystem and connect this idea to two major laws in science: the law of the conservation of mass and the law of the conservation of energy.

High School Students

High school students have sufficient knowledge of matter and energy to link these conversions to what happens in living and nonliving systems. Although there is no need to account for all the energy, students should know that most of the energy transferred from one organism to another is lost to the environment as heat and that this energy is not available for reuse. Knowledge of biogeochemical cycles is integrated across biology, chemistry, and Earth science, including the idea that some matter can remain in chemical reservoirs for a long time before it is cycled again.

Administering the Probe

This probe is most appropriate for middle and high school students. Make sure students know what a cycle is before using this probe.

Related Ideas in *National Science Education Standards* (NRC 1996)

K–4 Organisms and Their Environments

- All animals depend on plants. Some animals eat plants for food. Other animals eat animals that eat the plants.

5–8 Populations and Ecosystems

- For ecosystems, the major source of energy is sunlight. Energy entering ecosystems as sunlight is transferred by producers into chemical energy through photosynthesis. That energy then passes from organism to organism in food webs.

5–8 Structure of the Earth System

- Water, which covers the majority of the Earth's surface, circulates through the crust, oceans, and atmosphere in what is known as the "water cycle."

9–12 The Interdependence of Organisms

★ Energy flows through ecosystems in one direction, from photosynthetic organisms to herbivores to carnivores and decomposers.

9–12 Matter, Energy, and Organization in Living Systems

★ As matter and energy flows through different levels of organization of living systems—cells, organs, organisms, commu-

★ Indicates a strong match between the ideas elicited by the probe and a national standard's learning goal.

nities—and between living systems and the physical environment, chemical elements are recombined in different ways. Each recombination results in storage and dissipation of energy into the environment as heat. Matter and energy are conserved in each change.

Related Ideas in *Benchmarks for Science Literacy* (AAAS 2009)

K–2 Interdependence of Life

- Animals eat plants or other animals for food and may also use plants (or even other animals) for shelter and nesting.

3–5 Flow of Matter and Energy

- Almost all kinds of animals' food can be traced back to plants.
- Some source of "energy" is needed for all organisms to stay alive and grow.

6–8 Flow of Matter and Energy

- Energy can change from one form to another in living things.
- Organisms that eat plants break down the plant structures to produce the materials and energy they need to survive. Then they are consumed by other organisms.
- ★ Over a long time, matter is transferred from one organism to another repeatedly and between organisms and their physical environment. As in all material systems, the total amount of matter remains constant, even though its form and location change.

9–12 Flow of Matter and Energy

- ★ The chemical elements that make up the molecules of living things pass through food webs and are combined and recombined in different ways. At each link in a food web, some energy is stored in newly made structures but much is dissipated into the environment. Continual input of energy from sunlight keeps the process going.
- At times, environmental conditions are such that land and marine organisms reproduce and grow faster than they die and decompose to simple carbon-containing molecules that are returned to the environment. Over time, layers of energy-rich organic material inside the Earth have been chemically changed into great coal beds and oil pools.

Related Research

- Smith and Anderson (1986) found that almost all of the 12-year-old students in their sample were aware that there are cyclical processes in ecosystems. However, most of the students thought in terms of sequences of cause-and-effect events where matter was either created or destroyed and then the sequence repeated. They failed to recognize oxygen and carbon dioxide cycles or processes involving food. Their understanding of matter cycling was fragmented (Driver et al. 1994).
- Leach et al. (1992) found that some students fail to recognize that matter is conserved in the processes of photosynthesis, assimilation of food, decay, and respiration. Furthermore, they had difficulty distinguishing between food, matter, and energy.
- Gayford's (1986) study of 17- and 18-year-old biology students revealed that many students thought that energy flows from place to place and is stored like a material. They thought energy was either created or destroyed in biological processes rather than converted and conserved.
- Many children think that dead things simply disappear when they decay and are not cycled back into the environment (Driver et al. 1994).

★ Indicates a strong match between the ideas elicited by the probe and a national standard's learning goal.

Suggestions for Instruction and Assessment

- Combine this probe with "Earth's Mass" and "Rotting Apple" from *Uncovering Student Ideas in Science, Vol. 3: Another 25 Formative Assessment Probes* (Keeley, Eberle, and Dorsey 2008). Both of these probes address the concept of matter cycling by decomposers.
- Having students account for where the matter and energy go may help them see that matter can cycle from organisms to the environment and from the environment to organisms, but that energy flows only in one direction. Instruction that uses charts of the flow of matter through an ecosystem and emphasizes the reasoning involved with the entire process may enable students to develop more accurate conceptions (NRC 1996).
- Be sure to combine an understanding of matter cycling and energy flow with conservation of matter and energy in ecological cycles. However, be aware that the term *conserved* in an energy context can confuse students. Students may think conservation means the energy is still available for biological activity. Teachers should reiterate that the energy that is released as matter moves up the trophic pyramid is conserved—it does not simply disappear and cease to exist. However, organisms cannot hold onto that energy to do work. Rather, the energy is lost because it is converted into a form (heat) that most organisms cannot use efficiently. So even though *total* energy is conserved, the *useful* energy is lost from the working system.
- Understanding that the breakdown and reassembly of molecules during matter transformation needs to precede the idea of matter cycling. This is particularly important when using processes such as photosynthesis and cellular respiration to explain carbon and oxygen cycling.
- Combine the idea of *decomposers* with the concept of *recomposers*. This may help students recognize that these organisms not only break down matter into basic components but also make these components available for further use by organisms and the environment. However, make sure students know that *recomposer* is not a term used in biology; it is merely used to conceptualize what happens as a result of decomposition.

Related NSTA Science Store Publications, NSTA Journal Articles, NSTA SciGuides, NSTA SciPacks, and NSTA Science Objects

American Association for the Advancement of Science (AAAS). 2001. *Atlas of science literacy.* Vol. 1. (See "Flow of Matter in Ecosystems" map, pp. 76–77.) Washington, DC: AAAS.

American Association for the Advancement of Science (AAAS). 2001. *Atlas of science literacy.* Vol. 1. (See "Flow of Energy in Ecosystems" map, pp. 78–79.) Washington, DC: AAAS.

Koba, S., with A. Tweed. 2009. *Hard-to-teach biology concepts: A framework to deepen student understanding.* Arlington, VA: NSTA Press.

Trautmann, N. 2003. *Decay and renewal.* Arlington, VA: NSTA Press.

Related Curriculum Topic Study Guides (in Keeley 2005)
"Cycling of Matter in Ecosystems"
"Flow of Energy Through Ecosystems"

References

American Association for the Advancement of Science (AAAS). 2009. Benchmarks for science literacy online. *www.project2061.org/publications/bsl/online*

Driver, R., A. Squires, P. Rushworth, and V. Wood-Robinson. 1994. *Making sense of secondary science: Research into children's ideas.* London: RoutledgeFalmer.

Gayford, C. 1986. Some aspects of the problems of teaching about energy in school biology. *European Journal of Science Education* 8 (4): 443–450.

Keeley, P. 2005. *Science curriculum topic study: Bridging the gap between standards and practice.* Thousand Oaks, CA: Corwin Press and Arlington, VA: NSTA Press.

Keeley, P., F. Eberle, and C. Dorsey. 2008. *Uncovering student ideas in science, vol. 3: Another 25 formative assessment probes.* Arlington, VA: NSTA Press.

Leach, J., R. Driver, P. Scott, and C. Wood-Robinson. 1992. *Progression in conceptual understanding of ecological concepts by pupils age 5–16.* Leeds, UK: University of Leeds, Centre for Studies in Science and Mathematics Education.

National Research Council (NRC). 1996. *National science education standards.* Washington, DC: National Academies Press.

Smith, E., and C. Anderson. 1986. Alternative student conceptions of matter cycling in ecosystems. Paper presented to the National Association of Research in Science Teaching, San Francisco, California.

No More Plants

Four friends visited an island. The island was far away from the mainland. No humans lived on the island. The friends talked about what would happen if all the plants disappeared on the island. This is what they said:

Harold: "I think all the animals on the island would eventually die."

Jeff: "I think the animals that eat plants would eventually die but the animals that eat both plants and animals would live."

Salma: "I think only the predators on the island would live."

Misha: "I think eventually all the animals on the island will become meat eaters, and they will survive without plants."

Which friend do you agree with the most? ____________________ Explain why you agree.

____

____

____

____

____

____

____

No More Plants

Teacher Notes

Purpose

The purpose of this probe is to elicit students' ideas about the role of plants in a terrestrial ecosystem. The probe is designed to determine whether students recognize that all the animals depend on plants, whether or not they eat plants.

Related Concepts

Producer, consumer, food chain, food web, ecosystem, interdependence

Explanation

Harold has the best answer: "I think all the animals on the island would eventually die." All animals depend on plants or other photosynthetic organisms (such as phytoplankton) whether they eat them or not. Plants and other photosynthetic organisms are the basic source of energy for all animals. Plants use the energy from sunlight to make food that contains high-energy chemical bonds. When animals eat plants, this chemical energy is transferred to animals. Animals that eat other animals are also consuming food that contains energy traced back to plants or other photosynthetic organisms. Without plants or other photosynthetic organisms, there would be no food. In addition, plants and photosynthetic organisms such as phytoplankton provide oxygen as a by-product of photosynthesis. They take in carbon dioxide and water and, using energy from sunlight, manufacture sugars, releasing oxygen during the process. Oxygen is needed by all plants and animals for respiration. Without plants or other photosynthetic organisms, there would be no cycling of carbon and oxygen through living and nonliving components of an ecosystem. Animals simply could not survive without photosynthetic organisms.

Curricular and Instructional Considerations

Elementary Students

In the elementary grades, students learn that all animals depend on plants. They trace all

food back to plants using food chains and food webs. However, the details of photosynthesis and matter cycling and energy flow can wait until middle school.

Middle School Students

In the middle grades, students begin to develop a matter cycling and energy flow explanation for why animals depend on plants. They distinguish between producers and consumers in food webs and trace the matter back to plants and energy back to the Sun. As they examine multiple links in a food web, they should be able to trace back all links to plants.

High School Students

At the high school level, students build on basic ecological principles they learned in middle school. They examine the effect of removing a population from an ecosystem and the limitations of scarce resources when a community competes. Conceptual understandings of interdependence contribute to their growing understanding of contemporary ecological issues.

Administering the Probe

This probe is best used in grades K–8. Make sure students know that the island is geographically isolated and the animals cannot move off of the island to a new habitat.

Related Ideas in *National Science Education Standards* (NRC 1996)

K–4 Organisms and Their Environments

- ★ All animals depend on plants. Some animals eat plants for food. Other animals eat animals that eat the plants.

5–8 Populations and Ecosystems

- Populations of organisms can be categorized by the function they serve in an ecosystem. Plants and some microorganisms are producers—they make their own food. All animals, including humans, are consumers, which obtain food by eating other organisms. Decomposers, primarily bacteria and fungi, are consumers that use waste materials and dead organisms for food. Food webs identify the relationships among producers, consumers, and decomposers in an ecosystem.
- ★ For ecosystems, the major source of energy is sunlight. Energy entering ecosystems as sunlight is transferred by producers into chemical energy through photosynthesis. That energy then passes from organism to organism in food webs.

9–12 The Interdependence of Organisms

- ★ The atoms and molecules on the Earth cycle among the living and nonliving components of the biosphere.
- Organisms both cooperate and compete in ecosystems. The interrelationships and interdependencies of these organisms may generate ecosystems that are stable for hundreds or thousands of years.

Related Ideas in *Benchmarks for Science Literacy* (AAAS 2009)

K–2 Interdependence of Life

- Animals eat plants or other animals for food and may also use plants (or even other animals) for shelter and nesting.

3–5 Flow of Matter and Energy

- ★ Almost all kinds of animals' food can be traced back to plants.

6–8 Flow of Matter and Energy

- ★ Organisms that eat plants break down the plant structures to produce the materials

★ Indicates a strong match between the ideas elicited by the probe and a national standard's learning goal.

and energy they need to survive. Then they are consumed by other organisms.

- Almost all food energy originally comes from sunlight.

9–12 Flow of Matter and Energy

- The chemical elements that make up the molecules of living things pass through food webs and are combined and recombined in different ways. At each link in a food web, some energy is stored in newly made structures but much is dissipated into the environment. Continual input of energy from sunlight keeps the process going.

Related Research

- Some students believe that when there is a change in the population of one organism, it only affects the organisms that are adjacent to it in the food chain, such as primary consumers being the only ones affected when plants die (Allen 2010).
- A study conducted by Eisen and Stavy (1988) of students from age 13 to the undergraduate level revealed that most of the students knew that animals could not exist in a world without plants and suggested this was because animals cannot make their own food. Furthermore, some believed that carnivores could exist if their prey reproduced plentifully.
- Few students appear to relate their ideas about feeding and energy to a framework of ideas about interactions of organisms. Only about half of a sample of undergraduate biology students, when asked about the phrases "life depends on green plants" and "the web of life," explained these phrases in terms of food chains. Only a minority of these students mentioned harnessing solar energy or photosynthesis as the reason why plants are crucial in the food chain (Driver et al. 1994, p. 61).
- In a study by Griffiths and Grant (1985) of 15-year-olds, many of the students thought that a change in the population of one species would affect only those species related to it directly as predator or prey, while others thought a change in the prey population would have no effect on the predator population. The authors suggest that the introduction of food chain ideas as a prelude to food webs is a reason why children fail to use ideas about interdependency to explain relationships in complex ecosystems (Driver et al. 1994).

Suggestions for Instruction and Assessment

- For older students, combine this probe with the "Food Chain Energy" probe in this book.
- When introducing simple food chains, always trace the matter and energy back to plants (and energy back to the Sun). Repeat this when moving from food chains to food webs, making sure students know that a food chain is one part of a food web.
- Making the jump from simple linear food chains to more complex food chains and food webs can be problematic for students if students do not have an opportunity to see how linear food chains fit within a food web. Practice analyzing food chains and food webs will help students understand that these are interdependent systems. Making food webs by joining food chains together will help students see the interconnectedness of food chains.
- A food web wall using pictures and lengths of colored strings for connections can help students see how all the food that animals eat can ultimately be traced back to plants and the Sun.
- Have students explore other interdependencies by asking what would happen if all the predators disappeared or if all the pri-

mary consumers disappeared. Have them practice analyzing interdependencies in a food web by removing one organism from the web and describing how that would impact the rest of the food web.
- Ask older students what would happen in a marine ecosystem if all the phytoplankton suddenly died.

Related NSTA Science Store Publications, NSTA Journal Articles, NSTA SciGuides, NSTA SciPacks, and NSTA Science Objects

American Association for the Advancement of Science (AAAS). 2001. *Atlas of science literacy.* Vol. 1. (See "Flow of Matter in Ecosystems" map, pp. 76–77.) Washington, DC: AAAS.

American Association for the Advancement of Science (AAAS). 2001. *Atlas of science literacy.* Vol. 1. (See "Flow of Energy in Ecosystems" map, pp. 78–79.) Washington, DC: AAAS.

Koba, S., with A. Tweed. 2009. *Hard-to-teach biology concepts: A framework to deepen student understanding.* Arlington, VA: NSTA Press.

Related Curriculum Topic Study Guides (in Keeley 2005)
"Food Chains and Food Webs"
"Ecosystems"
"Interdependency Among Organisms"

References

Allen, M. 2010. *Misconceptions in primary science.* Berkshire, UK: Open University Press.

American Association for the Advancement of Science (AAAS). 2009. Benchmarks for science literacy online. *www.project2061.org/publications/bsl/online*

Driver, R., A. Squires, P. Rushworth, and V. Wood-Robinson. 1994. *Making sense of secondary science: Research into children's ideas.* London: RoutledgeFalmer.

Eisen, Y., and R. Stavy. 1988. Students' understanding of photosynthesis. *The American Biology Teacher* 50 (4): 208–212.

Griffiths, A., and B. Grant. 1985. High school students' understanding of food webs: Identification of a learning hierarchy and related misconceptions. *Journal of Research in Science Teaching* 22 (5): 421–426.

Keeley, P. 2005. *Science curriculum topic study: Bridging the gap between standards and practice.* Thousand Oaks, CA: Corwin Press and Arlington, VA: NSTA Press.

National Research Council (NRC). 1996. *National science education standards.* Washington, DC: National Academies Press.

Changing Environment

Two friends were talking about adaptations. They each had different ideas about what happens when an organism's environment changes so that it is very different from the organism's existing environment. This is what they said:

Leslie: "I don't think individual organisms can adapt to changes in their environments."

Jordan: "I think individual organisms can adapt to changes in their environments if they need to."

Whom do you most agree with? ____________________ Explain why you agree.

____

____

____

____

____

____

____

____

____

____

____

____

____

Changing Environment

Teacher Notes

Purpose

The purpose of this assessment probe is to elicit students' ideas about adaptation. The probe is designed to reveal whether students hold Lamarckian ideas about individual organisms adapting to changes in the environment.

Related Concepts

Adaptation, interdependence, natural selection

Explanation

Leslie has the best idea: "I don't think individual organisms can adapt to changes in their environments." In common usage, the word *adapt* is understood to mean any type of change over any span of time, whether it be during an organism's lifetime or the lifetimes of successive populations of organisms. However, in biology, when used in the context of natural selection and evolution, the word *adapt* refers to individuals that may have been born with a genetic variation that helps them survive in a changing environment. These individuals are then able to reproduce and pass on this trait to new generations so that their offspring will be *adapted* to the change. This is called natural selection. Adaptation generally is not intentional by an organism. The idea that an organism can pass on characteristics that it acquired during its lifetime to its offspring (also known as inheritance of acquired characteristics) was proposed by a French biologist named Jean-Baptiste Lamarck (1744–1829). Today, we know that organisms can adapt over time, but individuals do not intentionally adapt to a change during their lifetimes. Either they are born *adapted,* or they may die if they are not born with the genetic variation that will help them *adapt* to the change.

Some advanced students may have heard about or studied epigenetic phenomena, an exciting new branch of genetics research. These students may argue for both answers. Briefly, *epigenetics* refers to the study of heritable genetic changes that do not result from mutations in the DNA code but rather are due to changes in methylation or to other chemical modifications of DNA's backbone. These changes can appear in a single generation and occur in response to environmental stresses. More importantly, epigenetic changes in gene expression can be inherited by the next generation. Epigenetic inheritance initially seems to be a way for Lamarckian evolution to occur, because the environmental "experiences" of the parent generation can be imprinted onto their germ cell DNA and affect which genes are expressed in their offspring. However, epigenetically inherited traits are less common than Mendelian traits and are not subject to continuous natural selection. So even though it is possible for organisms to modify their genetic codes in response to environmental pressures (as Lamarck proposed), it is not a routine event and few students would recognize this.

Curricular and Instructional Considerations

Elementary Students

In the elementary grades, students build understandings of biological concepts through their direct experiences with living things and their habitats. The focus in the early elementary grades should be on establishing the primary association of organisms with their environments, followed by the development in the upper elementary grades of ideas about dependence on various aspects of the environment and the structures and behaviors that help organisms survive in that environment (NRC 1996). Students should have opportunities to investigate, preferably firsthand, a variety of plant and animal habitats and identify ways they depend on their environments and each other.

Middle School Students

Understanding *adaptation* is difficult at this level because of strongly held preconceptions students bring to their learning. Many middle school students think *adaptation* means that individuals intentionally change in major ways in response to major environmental changes. Middle school students need to understand the concept of variation and how organisms born with genetic variations that suit a major change in their environments are more apt to survive that change and reproduce than are organisms born without a variation that suits a major change. Those organisms that are more apt to survive and reproduce are also likely to pass on those genetic variations to their offspring. This is the level where students develop the fundamental ideas of natural selection that will help them later understand the mechanism of biological evolution.

High School Students

High school students progress from understanding an individual's response to a change in the environment to understanding a population's response. They shift from thinking about the selection of individuals with certain traits that help them survive in their environments to the changing proportion of such traits in a population of organisms. However, even at the high school level, some students will hold tenaciously to the traditional Lamarckian idea (as opposed to epigenetics) that individual organisms can adapt to environmental changes in ways that can be passed on to their offspring.

Administering the Probe

This probe can be used from upper elementary through high school. Make sure students know the environmental change referred to is a major change that affects the type of food available, shelter, and other needs an organism derives from its environment.

Related Ideas in *National Science Education Standards* (NRC 1996)

K–4 The Characteristics of Organisms

- Organisms have basic needs. For example, animals need air, water, and food; plants require air, water, nutrients, and light. Organisms can survive only in environments in which their needs can be met. The world has many different environments, and distinct environments support the life of different types of organisms.

K–4 Organisms and Their Environments

★ An organism's patterns of behavior are related to the nature of that organism's environment, including the kinds and numbers of other organisms present, the availability of food and resources, and the physical characteristics of the environment. When the environment changes, some plants and animals survive and reproduce, and others die or move to new locations.

5–8 Diversity and Adaptations of Organisms

★ Biological evolution accounts for the diversity of species developed through gradual processes over many generations. Species acquire many of their unique characteristics through biological adaptation, which involves the selection of naturally occurring variations in populations. Biological adaptations include changes in structures, behaviors, or physiology that enhance survival and reproductive success in a particular environment.

- Extinction of a species occurs when the environment changes and the adaptive characteristics of a species are insufficient to allow its survival. Fossils indicate that many organisms that lived long ago are extinct. Extinction of species is common; most of the species that have lived on the Earth no longer exist.

9–12 The Behavior of Organisms

- Like other aspects of an organism's biology, behaviors have evolved through natural selection. Behaviors often have an adaptive logic when viewed in terms of evolutionary principles.

9–12 Biological Evolution

- Species evolve over time. Evolution is the consequence of the interactions of (1) the potential for a species to increase its numbers, (2) the genetic variability of offspring due to mutation and recombination of genes, (3) a finite supply of the resources required for life, and (4) the ensuing selection by the environment of those offspring better able to survive and leave offspring.

Related Ideas in *Benchmarks for Science Literacy* (AAAS 2009)

K–2 Evolution of Life

- Different plants and animals have external features that help them thrive in different kinds of places.

K–2 Heredity

- There is variation among individuals of one kind within a population.

3–5 Evolution of Life

★ Individuals of the same kind differ in their characteristics, and sometimes the differences give individuals an advantage in surviving and reproducing.

★ Indicates a strong match between the ideas elicited by the probe and a national standard's learning goal.

3–5 Interdependence of Life

- For any particular environment, some kinds of plants and animals thrive, some do not live as well, and some do not survive at all.
- Changes in an organism's habitat are sometimes beneficial to it and sometimes harmful.

6–8 Evolution of Life

★ Individual organisms with certain traits are more likely than others to survive and have offspring.
★ Changes in environmental conditions can affect the survival of individual organisms and entire species.
- Extinction of species occurs when the environment changes and the individual organisms of that species do not have the traits necessary to survive and reproduce in the changed environment.

6–8 Interdependence of Life

- In any particular environment, the growth and survival of organisms depend on the physical conditions.

9–12 Evolution of Life

★ Natural selection provides the following mechanism for evolution: some variation in heritable characteristics exists within every species; some of these characteristics give individuals an advantage over others in surviving and reproducing; and the advantaged offspring, in turn, are more likely than others to survive and reproduce. As a result, the proportion of individuals that have advantageous characteristics will increase.
★ Heritable characteristics influence how likely an organism is to survive and reproduce.
- Natural selection leads to organisms that are well-suited for survival in particular environments.

★ When an environment, including other organisms that inhabit it, changes, the survival value of inherited characteristics may change.

6–8 Interdependence of Life

- Ecosystems can be reasonably stable over hundreds or thousands of years. As any population grows, its size is limited by one or more environmental factors: availability of food, availability of nesting sites, or number of predators.

Related Research

- Many students tend to see adaptation as an intention by the organism to satisfy a desire or need for survival. They tend to believe in the Lamarckian theory of inheritance of acquired characteristics. This belief is common both before and after instruction in genetics and evolution (Driver et al. 1994).
- An older study by Brumby (1979) of Australian and English biology students showed that even after studying upper-level biology, only 18% of the students could correctly apply natural selection to evolutionary change. Most believed that individuals can adapt to change in the environment if they need to (Lamarckian belief) and that these adaptations are then inherited.
- Middle and high school students may believe that organisms are able to intentionally change their bodily structures to be able to live in a particular habitat or that they respond to a changed environment by seeking a more favorable environment. It has been suggested that the terminology about the concept of adaptation used by teachers or textbooks may cause or reinforce these beliefs (AAAS 1993, p. 342).

Suggestions for Instruction and Assessment

- Combine this probe with "Habitat Change" in *Uncovering Student Ideas in Science, Vol. 2: 25 More Formative Assessment Probes*

★ Indicates a strong match between the ideas elicited by the probe and a national standard's learning goal.

(Keeley, Eberle, and Tugel 2007) or "Adaptation" and "Is It 'Fitter'?" in *Uncovering Student Ideas in Science, Vol. 4: 25 New Formative Assessment Probes* (Keeley and Tugel 2009).

- Probe deeper to find out if students who chose Jordan also believe those adaptations will be passed on to the organisms' offspring—a key belief of students who hold the Lamarckian view of acquired characteristics.
- When dealing with individual organisms, *acclimatization* would be a better term to use for noninheritable changes that some organisms are able to make during their lifetimes (e.g., humans put on a coat during cold weather).
- It has been suggested that Lamarckian interpretations of an individual's adaptation to its environment may impede understanding of Darwinian evolution. For this reason, it is important to use a variety of formative assessments to elicit students' preconceptions about adaptation.
- A common activity in elementary and middle school is to have students choose an existing organism (or invent an imaginary one) that is adapted to a particular habitat. Students are then presented with a change in the habitat and have to change features of the organism so it can survive in the changed habitat. Be aware that this activity may perpetuate the misconception that organisms intentionally adapt and that variations are not incremental changes over time but rather drastic, complete changes. A better activity would be to have the teacher choose five or six organisms with slight differences in characteristics and then ask which would survive best in varying environments.
- Have students compare and contrast the everyday use of the word *adapt* with the scientific meaning of the word. Add this to students' growing list of examples of the ways we use words in our everyday language that are not always the same in meaning when the word is used in a scientific context.
- It is important to help students understand that in most cases, animals do not intentionally change their physical structures, behavior, and niche in response to a major environmental change. However, in some cases plants and animals do change their body structures as part of a physiological response to environmental stress, a process called phenotypic plasticity. For example, a beaver in a cold climate will grow larger and have thicker fat and denser fur than a beaver in a warmer area. Yet if the cold-adapted animal is relocated to a warmer place, its body fat, hair growth, and feeding behaviors will change dramatically.

  Similarly, a plant's growth patterns can change as water availability, light levels, and soil nutrients change. In neither case is there an intentional change by the organism, but there is structural change nonetheless. To be sure, a beaver could not adapt to a marine environment, but it can change its own body quite dramatically. Similarly, many animals do respond behaviorally to a change in their environments by actively seeking more favorable conditions. Both local and long-distance migrations take place for this reason. To give another example, the distribution of fish species in the water column of a lake depends on many variables; if one variable such as water temperature changes so conditions are no longer favorable, individual fish can migrate up or down in the water column, or to another location entirely, to find a more "favorable" thermal environment.

Related NSTA Science Store Publications, NSTA Journal Articles, NSTA SciGuides, NSTA SciPacks, and NSTA Science Objects

American Association for the Advancement of Science (AAAS). 2001. *Atlas of science literacy.* Vol. 1. (See "Natural Selection" map, pp. 82–83.) Washington, DC: AAAS.

Biological Sciences Curriculum Study (BSCS). 2005. *The nature of science and the study of biological evolution.* Colorado Springs, CO: BSCS.

Diamond, J., with C. Zimmer, E. Evans, L. Allison, and S. Disbrow. 2006. *Virus and the whale: Exploring evolution in creatures large and small.* Arlington, VA: NSTA Press.

Koba, S., with A. Tweed. 2009. *Hard-to-teach biology concepts: A framework to deepen student understanding.* Arlington, VA: NSTA Press.

Sandro, L., J. Constible, and R. Lee. 2007. Extreme arthropods: Exploring evolutionary adaptations to polar and temperate deserts. *Science Scope* 30 (9): 24–32.

Related Curriculum Topic Study Guides (in Keeley 2005)
"Adaptation"
"Natural and Artificial Selection"

References

American Association for the Advancement of Science (AAAS). 1993. *Benchmarks for science literacy.* New York: Oxford University Press.

American Association for the Advancement of Science (AAAS). 2009. Benchmarks for science literacy online. *www.project2061.org/publications/bsl/online*

Brumby, M. 1979. Problems in learning the concepts of natural selection. *Journal of Biological Education* 13 (2): 119–122.

Driver, R., A. Squires, P. Rushworth, and V. Wood-Robinson. 1994. *Making sense of secondary science: Research into children's ideas.* London: RoutledgeFalmer.

Keeley, P. 2005. *Science curriculum topic study: Bridging the gap between standards and practice.* Thousand Oaks, CA: Corwin Press and Arlington, VA: NSTA Press.

Keeley, P. 2008. *Science formative assessment: 75 practical strategies for linking assessment, instruction, and learning.* Thousand Oaks, CA: Corwin Press and Arlington, VA: NSTA Press.

Keeley, P., F. Eberle, and J. Tugel. 2007. *Uncovering student ideas in science, vol. 3: 25 more formative assessment probes.* Arlington, VA: NSTA Press.

Keeley, P., and J. Tugel. 2009. *Uncovering student ideas in science, vol. 4: 25 new formative assessment probes.* Arlington, VA: NSTA Press

National Research Council (NRC). 1996. *National science education standards.* Washington, DC: National Academies Press.

Eggs

Chicks come from eggs. What other things come from eggs? Put an X next to each thing you think comes from an egg.

_____ whale
_____ worm
_____ rock
_____ alligator
_____ clam
_____ frog
_____ cat

_____ butterfly
_____ soil
_____ mouse
_____ robin
_____ catfish
_____ spider
_____ oak tree

_____ human
_____ bacteria
_____ beetle
_____ bean plant
_____ single-celled organism
_____ cow
_____ turtle

Explain your thinking. What rule or reasoning did you use to decide if something comes from an egg?

Eggs

Teacher Notes

Purpose

The purpose of this assessment probe is to elicit students' ideas about sexual reproduction. The probe is designed to reveal whether students recognize that multicellular organisms develop from a fertilized egg.

Related Concepts

Egg, sexual reproduction, life cycle

Explanation

The best answer is: Everything on the list comes from an egg except for four things—soil, bacteria, rock, and single-celled organism. The rest are multicellular plants and animals that reproduce sexually. During sexual reproduction, an egg cell is fertilized by a sperm cell. This fertilized egg then divides and develops into an organism. Soil and rock are not living things, so they do not come from eggs (although soil does contain living organisms). Although we often refer to eggs as animal in origin, biologically plants also produce eggs. At the base of a flower is the ovary. The ovary contains one or more ovules. Each ovule contains an egg that is fertilized by a sperm cell that develops in a pollen grain. The fertilized egg within the ovule develops into a seed, which may then germinate and develop into the plant.

Although this is a simplified explanation of a more complex process, the big idea is that both plants and animals develop from fertilized eggs. Only multicellular organisms can produce eggs. Bacteria do not produce eggs. Bacteria and other single-celled organisms generally reproduce asexually by binary fission or other means of cell division. Bacteria are single-celled, as are egg cells; therefore, a bacterium does not contain eggs within its cell. Viruses are not considered cells and are generally considered not to be living things. They can only reproduce by replicating within a living cell, using the cell's DNA. Rocks and soil are not living organisms and thus do not produce eggs, although tiny eggs of worms and insects can be found in soil.

Curricular and Instructional Considerations

Elementary Students

In the elementary grades, students learn about the life cycles of plants and animals. They observe the life cycles of various animals such as butterflies, frogs, and chicks, and learn that their life cycles begin with an egg. They learn that some animals lay their eggs and that these eggs develop outside the animal's body. They also learn that some animals develop inside the mother's body. They learn about the life cycles of plants and recognize that the life cycle of a plant begins with a seed. Elementary students would not be expected to understand that a seed developed from an egg within the ovule of a flower. Later in middle school, students learn about sexual reproduction in plants and recognize that seeds develop from a fertilized egg. See "Administering the Probe" for modifications appropriate for elementary students.

Middle School Students

At the middle school level, students learn about the difference between sexual and asexual reproduction. They learn that multicellular plants and animals develop from an egg fertilized by a sperm cell. They learn about the structure and function of different parts of a plant and animal, including the structures where eggs are produced. As they learn about the parts of a flower and the process of pollination, they recognize where the egg and sperm come from and what happens when the egg is fertilized.

High School Students

At the high school level, students build on their basic understanding of sexual reproduction in plants and animals and examine life cycles in more detail, including reproduction in flowering and nonflowering plants. Students use terms such as *gametophytes* to describe specialized cells involved in reproduction. The process of meiosis to produce eggs and sperm is emphasized at this level. However, some high school students will still revert to the common misconception that eggs come only from animals.

Administering the Probe

This probe is best used with students in grades 3–8. With younger children who are not yet ready to learn about plant reproduction, consider modifying the probe to include only animals. Younger students may only choose animals that lay their eggs outside of their bodies as egg producers, even though they learn that life cycles of animals begin with a fertilized egg. This probe can be administered as a card sort (Keeley 2008). Instead of a paper-pencil task, print each word from the list (p. 117) on cards. Give each small group a set of cards and have them sort them into two piles: things that produce eggs and things that do not produce eggs. Students discuss their justifications for each choice while they are doing the card sort. As you circulate among groups, listen carefully to students' reasoning and probe further if needed. For older elementary students, you might replace *bacteria* with *germ* or leave it out altogether.

Related Ideas in *National Science Education Standards* (NRC 1996)

K–4 Life Cycles of Organisms

- Plants and animals have life cycles that include being born, developing into adults, reproducing, and eventually dying. The details of this life cycle are different for different organisms.

5–8 Reproduction and Heredity

★ In many species, including humans, females produce eggs and males produce

★ Indicates a strong match between the ideas elicited by the probe and a national standard's learning goal.

sperm. Plants also reproduce sexually—the egg and sperm are produced in the flowers of flowering plants. An egg and sperm unite to begin development of a new individual. That new individual receives genetic information from its mother (via the egg) and its father (via the sperm). Sexually produced offspring never are identical to either of their parents.

9–12 Molecular Basis of Heredity

- Transmission of genetic information to offspring occurs through egg and sperm cells that contain only one representative from each chromosome pair. An egg and a sperm unite to form a new individual.

Related Ideas in *Benchmarks for Science Literacy* (AAAS 2009)

K–2 Human Development

- All kinds of animals have offspring, usually with two parents involved.

6–8 Heredity

★ In sexual reproduction, a single specialized cell from a female merges with a specialized cell from a male.

★ The fertilized egg cell, carrying genetic information from each parent, multiplies to form the complete organism.

6–8 Human Development

- Human fertilization occurs when sperm cells from a male's testes are deposited near an egg cell from the female ovary, and one of the sperm cells enters the egg cell.

Related Research

- Some children hold an "agricultural model" of reproduction, believing that eggs are "laid." Some student use this model to think of human eggs as similar to hens' eggs, incubated inside a mother's body (Driver et al. 1994).
- A transition in understanding reproduction happens after a child tries to make sense of the relationship between a mother and father, acquiring information about sexual intercourse and ideas about sperm and eggs. By age 11, most children understand the role of the parents' sperm and egg (Driver et al. 1994).
- In a study by Gott et al. (1985), 800 15-year-old students more often correctly identified sexual reproduction in animals than in plants. Many of the students did not believe that plants can reproduce sexually. Furthermore, studies confirmed by teachers showed that this view is very resistant to change, even after instruction. Biology instruction appears to have little effect in changing "folklore" concepts of reproduction (Driver et al. 1994).

Suggestions for Instruction and Assessment

- Combine this probe with "Does It Have a Life Cycle?" in *Uncovering Student Ideas in Science, Vol. 3: Another 25 Formative Assessment Probes* (Keeley, Eberle, and Dorsey 2008).
- When younger students learn about life cycles, starting with the egg, you should explicitly develop the idea that this stage in the life cycle is similar for all animals. All animals start out as an egg, even those that develop inside their mothers, such as mammals. Later, when students are ready to learn about the reproductive parts of a flower, you can extend this generalization to plants. Once students have learned about cells, they can understand that a life cycle starts with a single-celled egg.
- Probe deeper to find out what students' conceptions of an egg is inside animals that give live birth. Some students may

★ Indicates a strong match between the ideas elicited by the probe and a national standard's learning goal.

still hold the "agricultural model" of an egg with a developing fetus inside it, rather than a model of cell division starting with a fertilized egg.

- Have students compare and contrast the everyday common use of the word *egg* with the scientific meaning of the word. For example, in everyday usage, we think of an egg as something that is laid or has a shell. Add this to students' growing list of examples of the ways we use words in our everyday language that are not always the same in meaning when the word is used in a scientific context.

Related NSTA Science Store Publications, NSTA Journal Articles, NSTA SciGuides, NSTA SciPacks, and NSTA Science Objects

Koba, S., with A. Tweed. 2009. *Hard-to-teach biology concepts: A framework to deepen student understanding.* Arlington, VA: NSTA Press.

Konicek-Moran, R. 2009. Flowers are more than just pretty. In *More everyday science mysteries: Stories for inquiry-based science teaching,* 121–134. Arlington, VA: NSTA Press.

Science Object: *Cell Structure and Function: Cells—The Basis of Life*

Related Curriculum Topic Study Guide (in Keeley 2005)
"Reproduction, Growth, and Development (Life Cycles)"

References

American Association for the Advancement of Science (AAAS). 2009. Benchmarks for science literacy online. *www.project2061.org/publications/bsl/online*

Driver, R., A. Squires, P. Rushworth, and V. Wood-Robinson. 1994. *Making sense of secondary science: Research into children's ideas.* London: RoutledgeFalmer.

Gott, R., A. Davey, R. Gamble, J. Head, N. Khaligh, P. Murphy, T. Orgee, B. Schofield, and G. Welford. 1985. *Science in schools: Ages 13 and 15.* Report No. 3. London, UK: Department of Education and Science, Assessment and Performance Unit.

Keeley, P. 2005. *Science curriculum topic study: Bridging the gap between standards and practice.* Thousand Oaks, CA: Corwin Press and Arlington, VA: NSTA Press.

Keeley, P. 2008. *Science formative assessment: 75 practical strategies for linking assessment, instruction, and learning.* Thousand Oaks, CA: Corwin Press and Arlington, VA: NSTA Press.

Keeley, P., F. Eberle, and C. Dorsey. 2008. *Uncovering student ideas in science, vol. 3: Another 25 formative assessment probes.* Arlington, VA: NSTA Press.

National Research Council (NRC). 1996. *National science education standards.* Washington, DC: National Academies Press.

Chrysalis

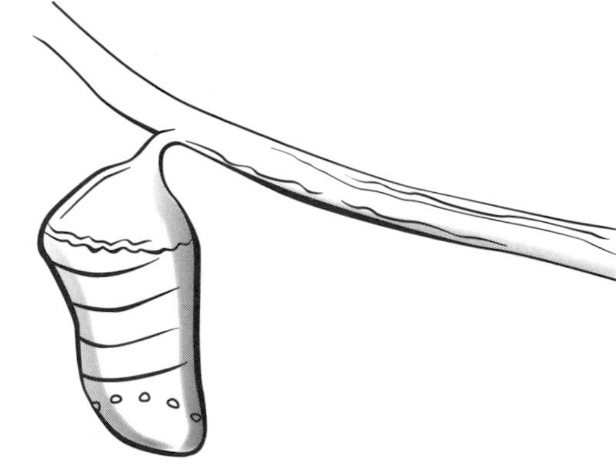

Three friends found a monarch butterfly caterpillar. They put it in a jar with a milkweed plant. The next day, they saw that the caterpillar had turned into a chrysalis. They wondered if the chrysalis was alive. This is what they said:

Mac: "I think the chrysalis is alive."

Lila: "I think the chrysalis is no longer a living organism."

Antoine: "I think the chrysalis is dead, but the butterfly that comes out is alive."

Which friend do you agree with the most? ____________ Explain why you agree.

____

____

____

____

____

____

____

____

____

____

____

____

____

Chrysalis

Teacher Notes

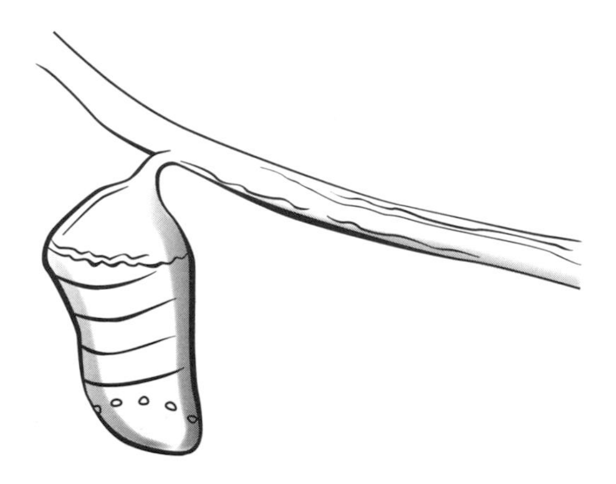

Purpose

The purpose of this assessment probe is to elicit students' ideas about life cycles. The probe is designed to see if students recognize that the pupal stage of a life cycle is a living organism.

Related Concepts

Life cycle, living, pupa, chrysalis

Explanation

The best answer is Mac's: "I think the chrysalis is alive." Some insects, such as butterflies, go through a complete metamorphosis during their life cycles in the stages from egg to larva to pupa to adult. The chrysalis is the pupal stage of a butterfly's life cycle. A life cycle is described by the stages an organism goes through during its life. During this stage, the larval structures are broken down and the adult structures of the butterfly are formed. During this amazing transformation, the organism is inactive (dormant) but it is still alive. Organisms cannot come back to life after death; therefore, logic also explains why Mac is correct and Antoine is incorrect.

Curricular and Instructional Considerations

Elementary Students

In the elementary grades, students learn about the life cycles of plants and animals. They observe the life cycles of various animals such as butterflies, frogs, and chicks, and learn that their life cycle begins with an egg and develops over time to become an adult. Stages of metamorphosis that involve larval and pupal stages are commonly addressed at the elementary level, with opportunities to observe the organisms' life cycles. As students progress through the elementary grades, they develop an understanding of several basic observable functions performed by living things, such as eating, drinking, breathing, growing, and moving. Because they do not observe a chrysalis doing any of these things, they may think it is nonliving.

Middle School Students

As students investigate a variety of life forms, they refine their earlier ideas about whether something is living. They recognize that some organisms go through a dormant period but that even during this period they still carry out the necessary functions to sustain life. Students also develop a deeper understanding of cycles and should recognize that a life cycle cannot continue into adulthood—when organisms reproduce and keep the cycle going—if an organism dies before reaching adulthood.

High School Students

By high school, students have a more complete understanding of the processes that support life, particularly at the cellular level. They have a greater ability to recognize the ubiquitous features and processes common to all life, from early stages of development through adulthood. By high school, they recognize death as the cessation of all life processes.

Administering the Probe

This probe is best used at the elementary level as an elicitation before studying the life cycle of a butterfly or for monitoring attainment of concepts after students have learned about a variety of life cycles. Consider showing students an image of a monarch chrysalis or the real thing before students respond to the probe.

Related Ideas in *National Science Education Standards* (NRC 1996)

K–4 Life Cycles of Organisms

★ Plants and animals have life cycles that include being born, developing into adults, reproducing, and eventually dying. The details of this life cycle are different for different organisms.

5–8 Structure and Function in Living Systems

- All organisms are composed of cells, the fundamental units of life.

9–12 Matter, Energy, and Organization in Living Systems

- All matter tends toward more disorganized states. Living systems require a continuous input of energy to maintain their chemical and physical organizations. With death, and the cessation of energy input, living systems rapidly disintegrate.
- As matter and energy flow through different levels of organization of living systems—cells, organs, organisms, communities—and between living systems and the physical environment, chemical elements are recombined in different ways. Each recombination results in storage and dissipation of energy into the environment as heat. Matter and energy are conserved in each change.

Related Ideas in *Benchmarks for Science Literacy* (AAAS 2009)

K–2 Flow of Matter and Energy

- Many materials can be recycled and used again, sometimes in different forms.

3–5 Flow of Matter and Energy

- Some source of "energy" is needed for all organisms to stay alive and grow.

6–8 Cells

- All living things are composed of cells, from just one to many millions, whose details usually are visible only through a microscope.

6–8 Flow of Matter and Energy

- Food provides molecules that serve as fuel and building material for all organisms.

★ Indicates a strong match between the ideas elicited by the probe and a national standard's learning goal.

Related Research

- In a study by Tamir, Gal-Chappin, and Nussnovitz (1981) of children ages 10 to 14, 19% did not understand the continuity of life. They believed that larvae changed into pupae, which are dead, and then they became butterflies.
- Students have various ideas about what constitutes living. Some may believe that things must be visibly eating and breathing to be alive. People of all ages tend to use movement, especially movement in response to a stimulus, as a characteristic of life. Young children frequently give "growth" as a criterion for life (Driver at al. 1994).
- Elementary and middle school students use observable processes such as movement, breathing, and reproducing when deciding whether something is alive or dead. High school and college students use these same readily observable characteristics to decide if something is alive. They rarely mention ideas such as "being made of cells" or biochemical details. It has been suggested that learning the facts that define life has contributed little toward understanding. Students may be able to quote the seven characteristics of life (moving, respiration, reaction to stimuli, growth, reproduction, elimination of waste, and need for food) but may not be able to apply them when determining if something is living (Brumby 1982).
- Carey (1985) suggested that progression in understanding the concept of *living* is linked to growth in children's ideas about biological processes. Young children have little knowledge of biology. In addition, it isn't until around the age of 9 or 10 that children begin to understand death as the cessation of life processes.

Suggestions for Instruction and Assessment

- Combine this probe with "Is It Living?" from *Uncovering Student Ideas in Science, Vol. 1: 25 Formative Assessment Probes* (Keeley, Eberle, and Farrin 2005), which further reveals students' ideas about whether eggs or pupae are living. It may also be helpful to combine this probe with the "Cucumber Seeds" probe found on page 9 of this book. That probe also deals with a dormant condition.
- Place emphasis on the "living" aspects when students are learning about life cycles and show that death is the end of the life cycle for an individual organism. Students often think that organisms in dormant or metamorphic states are dead. Counteract this misconception with what happens after these organisms emerge from such states and with the idea that they had to be alive for their life cycles to continue.
- Help older students distinguish between needs and processes. For example, you may not see the chrysalis eating, but it still needs energy to carry out the transformation of its body. That energy is supplied by the milkweed the caterpillar ate before it went into the pupal state. The molecules from that food are providing the energy the cells need to transform into a butterfly. You may not see the chrysalis breathing, but it still requires oxygen for respiration to get energy from its food.
- Ask students what it would take to determine whether or not the chrysalis is living. Use the mnemonic MRS GREN to help elementary students identify the seven life processes that characterize life: M = movement, R = respiration, S = stimuli (reaction to), G = growth, R = reproduction, E = elimination of wastes, and N = nutrition (needs

food). Make sure that students know what these processes mean and that not all living organisms will show all of them at the same time. If a time-lapse video of a chrysalis is available, show it to students so that they can see that a chrysalis is surprisingly active and even moves.

- The transformation going on in the chrysalis stage can be connected to the transformation of matter in living systems. The cells of the body structures of the caterpillar are being rearranged into the new structures of the butterfly using energy from food. Matter is conserved during the process.

Related NSTA Science Store Publications, NSTA Journal Articles, NSTA SciGuides, NSTA SciPacks, and NSTA Science Objects

Konicek-Moran, R. 2009. Oatmeal bugs. In *Everyday science mysteries: Stories for inquiry-based science teaching,* 89–97. Arlington, VA: NSTA Press.

Shimkanin, J., and A. Murphy. 2007. Let monarchs rule. *Science and Children* 45 (1): 32–36.

Related Curriculum Topic Study Guide (in Keeley 2005)
"Reproduction, Growth, and Development (Life Cycles)"

References

American Association for the Advancement of Science (AAAS). 2009. Benchmarks for science literacy online. *www.project2061.org/publications/bsl/online*

Brumby, M. 1982. Students' perceptions of the concept of life. *Science Education* 66 (4): 613–622.

Carey, S. 1985. *Conceptual change in childhood.* Cambridge, MA: MIT Press.

Driver, R., A. Squires, P. Rushworth, and V. Wood-Robinson. 1994. *Making sense of secondary science: Research into children's ideas.* London: RoutledgeFalmer.

Keeley, P. 2005. *Science curriculum topic study: Bridging the gap between standards and practice.* Thousand Oaks, CA: Corwin Press and Arlington, VA: NSTA Press.

Keeley, P., F. Eberle, and L. Farrin. 2005. *Uncovering student ideas in science, vol. 1: 25 formative assessment probes.* Arlington, VA: NSTA Press.

National Research Council (NRC). 1996. *National science education standards.* Washington, DC: National Academies Press.

Tamir, P., R. Gal-Chappin, and R. Nussnovitz. 1981. How do intermediate and junior high school students conceptualize living and nonliving? *Journal of Research in Science Teaching* 18 (3): 241–248.

DNA, Genes, and Chromosomes

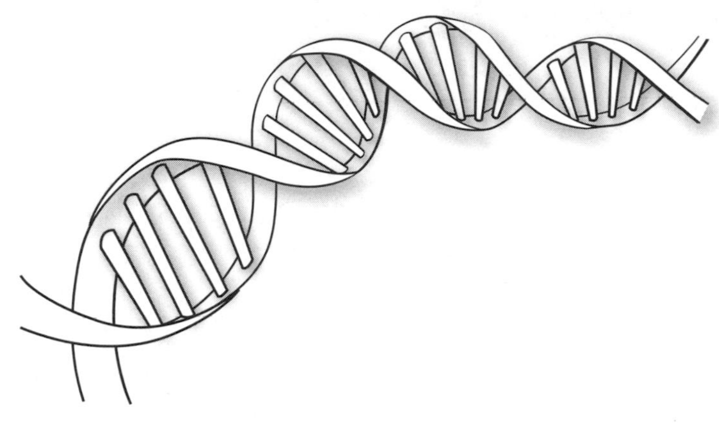

Four friends were talking about human DNA, genes, and chromosomes. They each had different ideas about where these structures were found. This is what they said.

Bella: "I think DNA is found on genes."

Kyra: "I think chromosomes are found on genes."

Kurt: "I think genes are found on DNA."

Joan: "I think chromosomes are found on DNA."

Whom do you agree with the most? __________ Explain why you agree.

____

____

____

____

____

____

____

DNA, Genes, and Chromosomes

Teacher Notes

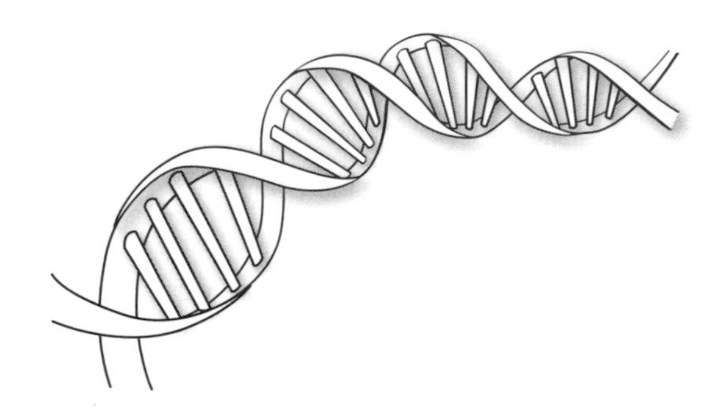

Purpose

The purpose of this assessment probe is to elicit students' ideas about structures they encounter when they learn about heredity. The probe is designed to reveal students' ideas about the "parts and wholes" relationship between DNA, genes, and chromosomes.

Related Concepts

DNA, genes, chromosome

Explanation

The best answer is Kurt's: "I think genes are found on DNA." Genes are the basic structural and functional unit of heredity. They are found on chromosomes, which are made up of DNA, histones, and other support proteins; therefore, genes are found on DNA. A gene is a segment of DNA that has a specific location on a chromosome. Humans have 23 pairs of chromosomes. Our chromosomes come in pairs (except for the X and Y chromosome in men), and each pair is made up of a single molecule of double-stranded DNA tightly coiled many times around a protein called a histone that supports its structure. If you imagine this strand of DNA being uncoiled and stretched out, it might look like a long ladder. Each of the rungs of the ladder is called a base pair, and segments of these base pairs of varying lengths are called genes. Each gene contains a piece of genetic information that tells the cell to make a specific protein. Thousands of genes are found on each strand of DNA that makes up your chromosomes.

It had been thought that much of the length of DNA does not seem to code for any specific protein and does not seem to be genes. This was long referred to as "junk DNA" and is now more often referred to as noncoding and structural DNA. Current evidence indicates that these regions are important for regulating gene activity. They also help position DNA in three-dimensional space within the nucleus, which in turn affects rates of gene expression. Put another way, DNA has information stored in its folding patterns, not just in the GATC code.

Curricular and Instructional Considerations

Elementary Students

At the elementary level, students develop basic ideas about heredity. The mechanism, cells, and molecular structures involved in inheriting genetic traits are not addressed until middle and high school.

Middle School Students

Middle school students begin to develop the genetic explanation for how traits are passed on from one generation to the next. They combine their basic understanding of genetics with their basic knowledge of cell structure, recognizing that chromosomes are found in the nucleus and genes are found on chromosomes. They learn that DNA is responsible for our inherited traits passed on by genes, but details about DNA should wait until students have a deeper understanding of molecules.

High School Students

In high school, students develop a deeper understanding of the structures involved in genetics at a molecular level and understand that genes code for specific proteins. They should be able to distinguish between DNA, chromosomes, and genes and know their relationship to one another. A growing understanding of biochemistry enables students to learn about the structure of the DNA molecule and how it functions in passing on inheritable traits. They learn about the pioneering studies of DNA and the scientific advances that led to the sequencing of the human genome.

Administering the Probe

This probe is best used at the middle and high school level with students who have some familiarity with these genetics concepts. At the high school level, it can be used to elicit students' preconceptions about the relationship among DNA, genes, and chromosomes. It can also be used to monitor students' understanding of the relationship among these three structures before proceeding with more advanced topics. Make sure students know that the DNA referred to in the answer choices is the complete strand of DNA, not a segment of DNA.

Related Ideas in *National Science Education Standards* (NRC 1996)

K–4 Life Cycles of Organisms

- Many characteristics of an organism are inherited from the parents of the organism, but other characteristics result from an individual's interaction with the environment.

5–8 Reproduction and Heredity

★ Hereditary information is contained in genes, located in the chromosomes of each cell. Each gene carries a single unit of information. An inherited trait of an individual can be determined by one or by many genes, and a single gene can influence more than one trait. A human cell contains many thousands of different genes.

9–12 The Molecular Basis of Heredity

★ In all organisms, the instructions for specifying the characteristics of the organism are carried in DNA, a large polymer formed from subunits of four kinds (A, G, C, and T). The chemical and structural properties of DNA explain how the genetic information that underlies heredity is both encoded in genes (as a string of molecular "letters") and replicated (by a templating mechanism). Each DNA molecule in a cell forms a single chromosome.

★ Indicates a strong match between the ideas elicited by the probe and a national standard's learning goal.

Related Ideas in *Benchmarks for Science Literacy* (AAAS 2009)

3–5 Heredity

- For offspring to resemble their parents, there must be a reliable way to transfer information from one generation to the next.

6–8 Heredity

- In organisms that have two sexes, typically half of the genes come from each parent.
- The same genetic information is copied in each cell of the new organism.

9–12 Heredity

★ Genes are segments of DNA molecules. Inserting, deleting, or substituting segments of DNA molecules can alter genes. An altered gene may be passed on to every cell that develops from it. The resulting features may help, harm, or have little or no effect on the offspring's success in its environment.

Related Research

- In his research on common misconceptions related to heredity, Berthelsen (1999) found that many students do not understand the relationship among DNA, genes, and chromosomes.
- Initial field test results of this probe show that some middle and high school students believe there is a hierarchical organization inside the nucleus in which the chromosome is the structural unit that contains genes and DNA molecules are found within these genes.
- A study of 16-year-old students in England and Wales showed a poor understanding of the processes by which genetic information is transferred and a lack of basic knowledge about the structures involved. There appeared to be widespread uncertainty and confusion over the use of genetics concepts such as genes and chromosomes (Lewis and Wood-Robinson 2000).

Suggestions for Instruction and Assessment

- Ask students what they would see if they had the magnification to see the details inside the nucleus of a cell. Ask them to keep zooming deeper from chromosome to DNA to the segments of DNA (genes) to the chemical bases that make up the genes.
- Take students on a "tour" of a chromosome. The Genetics Learning Center at the University of Utah has a nice example of an online tour at *http://learn.genetics.utah.edu/content/begin/traits/tour_chromosome.html.*
- Listen carefully when students choose "genes are found on DNA" as the answer in the probe. These students may not realize genes are segments with a specific location on a DNA molecule. They may think genes are just structures peppered throughout a DNA molecule. These students may also confuse each of the bases that make up a DNA molecule with being a gene.

Related NSTA Science Store Publications, NSTA Journal Articles, NSTA SciGuides, NSTA SciPacks, and NSTA Science Objects

American Association for the Advancement of Science (AAAS). 2001. *Atlas of science literacy.* Vol. 1. (See "DNA and Inherited Characteristics" map, pp. 68–69.) Washington, DC: AAAS.

Koba, S., with A. Tweed. 2009. *Hard-to-teach biology concepts: A framework to deepen student understanding.* Arlington, VA: NSTA Press.

Science Object: *Cell Structure and Function: The Molecular Machinery of Life.*

★ Indicates a strong match between the ideas elicited by the probe and a national standard's learning goal.

Related Curriculum Topic Study Guides (in Keeley 2005)
"DNA"
"Mechanism of Inheritance (Genetics)"

References

American Association for the Advancement of Science (AAAS). 2009. Benchmarks for science literacy online. *www.project2061.org/publications/bsl/online*

Berthelsen, B. 1999. Students' naïve conceptions in life science. *MSTA Journal* 44 (1): 13–19.

Keeley, P. 2005. *Science curriculum topic study: Bridging the gap between standards and practice.* Thousand Oaks, CA: Corwin Press and Arlington, VA: NSTA Press.

Lewis, J., and V. Wood-Robinson. 2000. Genes, chromosomes, cell division, and inheritance: Do students see any relationship? *International Journal of Science Education* 22 (2): 177–195.

National Research Council (NRC). 1996. *National science education standards.* Washington, DC: National Academies Press.

Eye Color

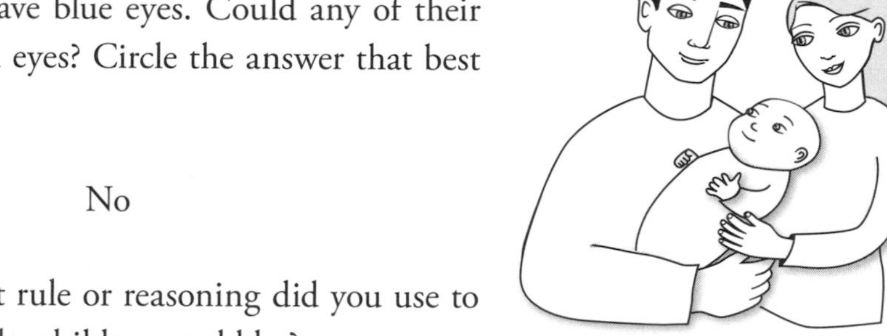

Mr. and Mrs. Miller both have blue eyes. Could any of their children be born with brown eyes? Circle the answer that best matches your thinking.

Yes No

Explain your thinking. What rule or reasoning did you use to decide what the eye color of the children could be?

Eye Color

Teacher Notes

Purpose

The purpose of this assessment probe is to elicit ideas about genetic traits. The probe is designed to see if students recognize that some traits, such as eye color, are complex and cannot be predicted solely by the result of one gene.

Related Concepts

Genes, traits

Explanation

The best answer is yes. Although the probability of having a blue-eyed child is the greatest, blue-eyed parents occasionally produce a brown-eyed child. It is a common misconception that only blue-eyed offspring result if both parents have blue eyes. This misconception often arises when students use Punnett squares to predict the eye color of offspring when parents have genes for either brown eyes or blue eyes. Mendelian genetics leads students to think that brown eye color is always dominant over blue eye color, which means (falsely) that two blue-eyed parents always have blue-eyed children. Eye color is a complex, polygenic (more than one gene) trait that depends on the state of interacting genes. Like many traits, eye color is not simply the result of one gene; other genes and factors that affect genes may also play a role in determining eye color. The genetics of eye color is quite complex and not entirely understood. However, the big idea is that eye color cannot be simply predicted by single dominant or recessive genes.

Curricular and Instructional Considerations

Elementary Students

In the elementary grades, students are beginning to learn about inherited characteristics versus characteristics that come from interactions with the environment or are learned behaviors. They observe that offspring do not always look exactly like their parents. In upper elementary grades students learn that parents

pass characteristics on to their offspring, but at this level it is still too early to introduce formal genetics concepts.

Middle School Students

In middle school, students learn basic ideas about the mechanisms of inheritance, cell division, and sexual reproduction. They are introduced to genetics and the role of genes in passing inherited traits from parents to offspring and the reasons why offspring may or may not resemble their parents. Students are frequently introduced to Punnett squares in middle school to predict gene combinations.

High School Students

In high school, students learn details about the mechanism of genetics, including at the molecular level, often with an emphasis on Mendelian genetics. They learn how various gene combinations, both single and polygenic, occur and how they express themselves. They learn that some traits, such as eye color, are complex and not easily predicted and may depend on the interaction of genes as well as on environmental factors.

Administering the Probe

This probe is best used at high school and is particularly useful to determine whether students have developed misconceptions based on prior learning about Mendelian genetics in middle school. The probe should not be given to students who are still struggling or have not yet been taught the basic Punnett square concepts of inheritance probabilities.

Related Ideas in *National Science Education Standards* (NRC 1996)

K–4 Life Cycles of Organisms

- Many characteristics of an organism are inherited from the parents of the organism, but other characteristics result from an individual's interactions with the environment.

5–8 Reproduction and Heredity

★ Hereditary information is contained in genes, located in the chromosomes of each cell. Each gene carries a single unit of information. An inherited trait of an individual can be determined by one or by many genes, and a single gene can influence more than one trait. A human cell contains many thousands of different genes.

- The characteristics of an organism can be described in terms of a combination of traits. Some traits are inherited and others result from interactions with the environment.

9–12 The Molecular Basis of Heredity

- Most of the cells in a human contain two copies of each of 22 different chromosomes. In addition, there is a pair of chromosomes that determines sex: a female contains two X chromosomes and a male contains one X and one Y chromosome. Transmission of genetic information to offspring occurs through egg and sperm cells that contain only one representative from each chromosome pair. An egg and a sperm unite to form a new individual. The fact that the human body is formed from cells that contain two copies of each chromosome—and therefore two copies of each gene—explains many features of human heredity, such as how variations that are hidden in one generation can be expressed in the next.

Related Ideas in *Benchmarks for Science Literacy* (AAAS 2009)

3–5 Heredity

- For offspring to resemble their parents, there must be a reliable way to trans-

★ Indicates a strong match between the ideas elicited by the probe and a national standard's learning goal.

fer information from one generation to the next.

6–8 Heredity

- In organisms that have two sexes, typically half of the genes come from each parent.

9–12 Heredity

★ The sorting and recombination of genes in sexual reproduction results in a great variety of possible gene combinations in the offspring of any two parents.
- The information passed from parents to offspring is coded in DNA molecules, long chains linking just four kinds of smaller molecules, whose precise sequence encodes genetic information.

Related Research

- A study conducted by the National Institutes of Health on high school students' essays from a genetics essay contest revealed several genetics misconceptions. One of these is that students believe one gene is always responsible for one trait (Shaw et al. 2008).
- The standard use of Punnett squares to solve Mendelian genetics problems contributes to students' misunderstandings of phenotypes and the complexities of the mechanism of inheritance (Moll and Allen 1987).

Suggestions for Instruction and Assessment

- The Stanford School of Medicine has a nice web page for public understanding of genetics. It features a geneticist explaining eye color to those who may lack the background knowledge to understand genetics (including why two blue-eyed parents can have a brown-eyed child). This site may be useful in explaining eye color genetics to students who have not yet developed a deep understanding of genetics principles: *www.thetech.org/genetics/ask.php?id=101.*
- Combine this probe with "Baby Mice" in *Uncovering Student Ideas in Science, Vol. 2: 25 More Formative Assessment Probes* (Keeley, Eberle, and Tugel 2007) to reveal further misconceptions about the concept of dominant genes.
- Caution should be used when students are asked to develop or use models to represent the mechanism of inheritance and predict the occurrence of traits. Some models, such as Punnett squares, oversimplify the occurrence of traits and fail to help students recognize that some traits, such as eye color, are polygenic.
- When planning instruction, be aware of the limitations of using Punnett squares. "The Punnett square can be used for studying the inheritance of genetic traits controlled by a single gene, and can even be applied when two or more traits are considered simultaneously, as long as the genes are not located on the same chromosome (linked). Students often learn to use Punnett squares to obtain correct answers to genetics problems, but they fail to understand that a Punnett square represents two biological processes—gamete formation and fertilization. Students rely on Punnett squares as algorithms for getting the 'right answer,' often at the expense of meaningful conceptual understanding" (Bryant 2003, p. 11).
- An interesting resource that can be used to show the complexity of determining genetic traits is a horse-coat-color genetics chart: *www.vgl.ucdavis.edu/services/coatcolorhorse.php.*

Related NSTA Science Store Publications, NSTA Journal Articles, NSTA SciGuides, NSTA SciPacks, and NSTA Science Objects

Bryant, R. 2003. Toothpick chromosomes: Simple manipulatives to help students understand genetics. *Science Scope* 26 (7): 10–15.

Koba, S., with A. Tweed. 2009. *Hard-to-teach biology concepts: A framework to deepen student understanding.* Arlington, VA: NSTA Press.

Rice, E., M. Krasny, and M. Smith. 2009. *Garden genetics: Teaching with edible plants.* Arlington, VA: NSTA Press.

Southworth, M., J. Mokros, C. Dorsey, and R. Smith. 2010. The case for cyberlearning: Genomics (and dragons!) in the high school biology classroom. *The Science Teacher* 77 (7): 28–33.

Related Curriculum Topic Study Guide (in Keeley 2005)
"Mechanism of Inheritance (Genetics)"

References

American Association for the Advancement of Science (AAAS). 2009. Benchmarks for science literacy online. *www.project2061.org/publications/bsl/online*

Bryant, R. 2003. Toothpick chromosomes: Simple manipulatives to help students understand genetics. *Science Scope* 26 (7): 10–15.

Keeley, P. 2005. *Science curriculum topic study: Bridging the gap between standards and practice.* Thousand Oaks, CA: Corwin Press and Arlington, VA: NSTA Press.

Moll, M., and R. Allen. 1987. Student difficulties with Mendelian genetics problems. *The American Biology Teacher* 49 (4): 229–233.

National Research Council (NRC). 1996. *National science education standards.* Washington, DC: National Academies Press.

Shaw, K., K. Horne, H. Zhang, and J. Boughman. 2008. Essay contest reveals misconceptions of high school students in genetics content. *Genetics* 178 (3): 1157–1168. *www.ncbi.nlm.nih.gov/pmc/articles/PMC2278104*

Human Body

Three students made different statements about cells and the human body. This is what they said:

Herman: "The human body is made up entirely of a collection of trillions of cells and things made from cells."

Felix: "The human body is surrounded by an outer covering. Inside this covering the body is filled in with cells and things made from cells."

Diandra: "The human body is a collection of trillions of cells and things made from cells contained inside an outer covering and inner coverings that contain organs."

Which student do you think best describes the human body? ________________.

Explain your thinking.

Human Body

Teacher Notes

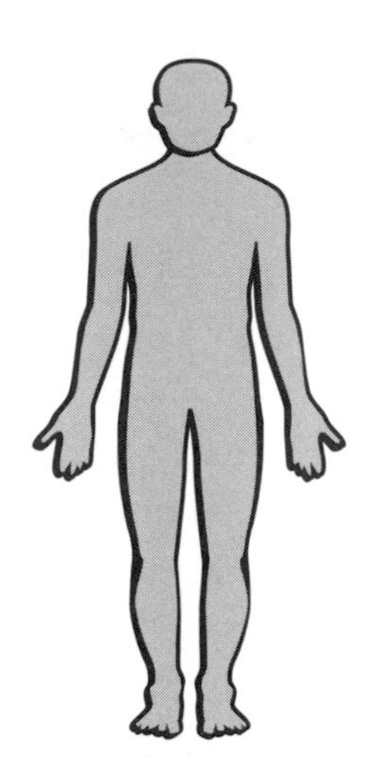

Purpose

The purpose of this assessment probe is to elicit students' ideas about the cellular makeup of the human body. The probe is designed to see if students recognize that the body is an organized collection of cells and not a structure or "outline" that contains or is filled in with cells.

Related Concepts

Human body, cells

Explanation

The best answer is Herman's: "The human body is made up entirely of a collection of trillions of cells and things made from cells." The human body is a complex system of cells that starts with a single, fertilized cell that divides to form an adult human organism made up of trillions of cells. Some students may think the body contains cells, rather than being cellular in composition (and containing cell secretions).

Students who select Diandra may draw on experiences where they learned that blood cells travel through the veins and arteries and that these structures contain cells inside of them. What they fail to realize is that veins and arteries are also made up of cells that form vascular tissue. Some students have similar conceptions of organs inside the human body. They may think organs are "sacs" filled with cells that perform a special function. Likewise, some students do not recognize skin as being cellular. They may think skin covers the body and that contained inside the skin are structures made of cells. Students who select Felix may have a conception of the body as a sac filled in with cells and cellular material.

Curricular and Instructional Considerations

Elementary Students

At the elementary level, younger students learn about the parts of the body that are visible to

them. In the intermediate grades, students learn about internal structures, with an emphasis on major body organs. Students may be introduced to the cells, sometimes with the phrase "the human body contains cells," which may give rise to an early conception that the body is like a "sac" filled with organs and cells, rather than with the clearer phrase, "the human body is made up of cells."

Middle School Students

At the middle level, students learn about the hierarchical arrangement of the human body, including the idea that the cell is the basic unit of both function and structure in the human body. Although students understand that the body is made up of cells, they may not understand that the body is cellular, rather than filled with cells.

High School Students

In high school, students deepen their understanding of cells, including the structures that make up cells and the molecules that make up cell structures. However, they may still conceptualize the human body as a structure that contains cells rather than being made up of cells.

Administering the Probe

This probe is best used with upper elementary and middle school students. Make sure students know that the probe is asking what the body is made of. It does not include things like the food and waste products inside the body that do not become a part of the structures that make up the body. You may wish to clarify the distinction between the three statements. The first statement means the body is cellular—it is composed entirely of cells or materials that come from cells, such as fingernails. The second statement means the body has a covering inside of which there are cells, as well as internal sacs, such as organs, that have cells inside of them. The covering is not considered to be entirely made up of cells or cell materials. The third statement means that the body has a covering that is not considered to be entirely made up of cells or cell materials, and that filling the inside of the body are cells and materials that are not made of cells.

Related Ideas in *National Science Education Standards* (NRC 1996)

5–8 Structure and Function in Living Systems

★ All organisms are composed of cells—the fundamental unit of life. Most organisms are single cells; other organisms, including humans, are multicellular.

9–12 The Cell

- Cells can differentiate, and complex multicellular organisms are formed as a highly organized arrangement of differentiated cells.

Related Ideas in *Benchmarks for Science Literacy* (AAAS 2009)

3–5 The Cell

- Microscopes make it possible to see that living things are made mostly of cells.

6–8 The Cell

★ All living things are composed of cells, from just one to many millions, whose details usually are visible only through a microscope.

- Different body tissues and organs are made up of different kinds of cells.

Related Research

- Dreyfus and Jungwirth's (1988) research showed that even though students are

★ Indicates a strong match between the ideas elicited by the probe and a national standard's learning goal.

introduced to the cell as the basic building block of life, many believe that cells are inside the body or that the body "contains" cells, rather than that the body is composed of cells.

Suggestions for Instruction and Assessment

- This probe can be combined with "Human Body Basics" from *Uncovering Student Ideas in Science, Vol. 1: 25 Formative Assessment Probes,* which addresses the idea that the cell is the basic unit of both structure and function in the human body (Keeley, Eberle, and Farrin 2005).
- Be aware that providing outlines of the human body and having students draw in the organs may contribute to students assuming that the human body is a sac filled with organs that contain cells.
- When students learn about parts of the body such as tendons and ligaments, point out that these structures are formed partly of cells and of cell secretions such as collagen and fibrin. Ask students where hair and fingernails come from. Point out that structures like hair and fingernails, although not composed of cells, arose from and were made from cell proteins.

Related NSTA Science Store Publications, NSTA Journal Articles, NSTA SciGuides, NSTA SciPacks, and NSTA Science Objects

American Association for the Advancement of Science (AAAS). 2001. *Atlas of science literacy.* Vol. 1. (See "Cell and Organs" map, pp. 74–75.) Washington, DC: AAAS.

Science Object: *Cell Structure and Function: The Basis of Life*

Related Curriculum Topic Study Guides (in Keeley 2005)
"Cells"
"Human Body Systems"

References

American Association for the Advancement of Science (AAAS). 2009. Benchmarks for science literacy online. *www.project2061.org/publications/bsl/online*

Dreyfus, A., and E. Jungwirth. 1988. The cell concept of 10th graders: Curricular expectations and reality. *International Journal of Science Education* 10: 221–229.

Keeley, P. 2005. *Science curriculum topic study: Bridging the gap between standards and practice.* Thousand Oaks, CA: Corwin Press and Arlington, VA: NSTA Press.

Keeley, P., F. Eberle, and L. Farrin. 2005. *Uncovering student ideas in science, vol. 1: 25 formative assessment probes.* Arlington, VA: NSTA Press.

National Research Council (NRC). 1996. *National science education standards.* Washington, DC: National Academies Press.

Human Excretory System

The excretory system is one of the important systems of the human body. Put an X next to all the things that are examples of things performed by the excretory system.

_____ vomiting

_____ eliminating feces (defecating)

_____ sweating

_____ producing saliva

_____ producing tears

_____ exhaling carbon dioxide

_____ urinating

_____ sneezing

Explain your thinking. What rule or reasoning did you use to decide if something on the list was performed by the excretory system?

___

___

___

___

Human Excretory System

Teacher Notes

Purpose

The purpose of this assessment probe is to elicit students' ideas about the human excretory system. The probe is designed to see if students recognize that the human excretory system removes metabolic wastes, rather than undigested food or other nonmetabolic excretions and secretions. The probe is also helpful in determining whether students distinguish between waste products produced through metabolism or waste products that result from undigested food (nonmetabolic).

Related Concepts

Digestive system, excretion, excretory system, metabolic waste, secretion

Explanation

The best answers are sweating, exhaling carbon dioxide, and urinating. The excretory system is responsible for getting rid of metabolic wastes that result from chemical reactions within the body. Urine from the kidneys contains urea (made by the liver to dispose of excess nitrogen), excess hydrogen, and a variety of unneeded salts. Perspiration eliminates metabolic wastes such as urea, lactic acid, ammonia, and salts. Most of the salt ions (sodium, potassium, chloride, magnesium, and calcium) in urine and perspiration are not true metabolic wastes because they came into the body from food; they are not created by metabolic processes. Exhaled air contains carbon dioxide, a waste product of cellular respiration, plus water vapor.

Excretion, as it is used with *the excretory system*, does not refer to elimination of solid waste from undigested food. Many people misuse the term *excretion*, confusing it with *defecation*. Even the word *excrement* sounds as if it is related to the excretory system. Excrement, or feces, are made up of water (about 70%), undigested food (about 20%), and gut bacteria (about 10%). Unlike urine, sweat, or exhaled air, very little of feces comes from within cells or are waste products of metabolic processes.

These wastes pass through the digestive system in a process correctly described in biological terms as *egestion*, rather than excretion. A few of the body's metabolic wastes that are secreted into the bile and then eliminated in the feces are considered to be excreted.

Vomit is food that is expelled before it goes through the digestive process. Saliva and tears are useful substances produced by cells. They are not waste products. Sneezing is a response to an irritating agent and may expel mucous, which is also a useful substance produced by cells to trap harmful substances.

Curricular and Instructional Considerations

Elementary Students

At the elementary level, students learn that all animals produce waste products, but the actual systems that are responsible for removing wastes are addressed at the middle school level.

Middle School Students

Middle school students learn that getting rid of wastes is one of the essential life processes. They identify different types of waste products. They learn about different body systems and their functions, including the excretory system and the digestive system. However, many middle school students refer to excretion as both urinary and digestive waste products and often lump both under the excretory system.

High School Students

At the high school level, students develop an understanding of the essential life process of getting rid of wastes at the cellular level and recognize that cells must get rid of the waste products of their metabolism. They can distinguish between metabolic wastes the body gets rid of through the excretory system (excretion) versus the undigested food wastes the body gets rid of through the digestive system (egestion).

Administering the Probe

This probe is best used with middle and high school students. Be aware that some students may be uncomfortable discussing the topic of "wastes" using terms like *feces* and *defecation*.

Related Ideas in *National Science Education Standards* (NRC 1996)

K–4 The Characteristics of Organisms

- Each plant or animal has different structures that serve different functions in growth, survival, and reproduction.

5–8 Structure and Function in Living Systems

★ The human organism has systems for digestion, respiration, reproduction, circulation, excretion, movement, control, and coordination, as well as for protection.

9–12 The Cell

- Cells have particular structures that underlie their functions. Every cell is surrounded by a membrane that separates it from the outside world. Inside the cell is a concentrated mixture of thousands of different molecules that form a variety of specialized structures that carry out such cell functions as energy production, transporting molecules, disposing waste, synthesizing new molecules, and storing genetic material.

Related Ideas in *Benchmarks for Science Literacy* (AAAS 2009)

3–5 Basic Functions

- The indigestible parts of food are eliminated.

★ Indicates a strong match between the ideas elicited by the probe and a national standard's learning goal.

6–8 Basic Functions

- ★ To burn food for the release of energy stored in it, oxygen must be supplied to cells, and carbon dioxide removed. Lungs take in oxygen for the combustion of food and eliminate the carbon dioxide produced. The urinary system disposes of dissolved waste molecules, the intestinal tract removes solid wastes, and the skin and lungs aid in the transfer of thermal energy from the body. The circulatory system moves all these substances to or from cells where they are needed or produced, responding to changing demands.
- For the body to use food for energy and building materials, the food must first be digested into molecules that are absorbed and transported to cells.

6–8 The Cell

- Various organs and tissues function to serve the needs of all cells for food, air, and waste removal.
- ★ Within cells, many of the basic functions of organisms—such as extracting energy from food and getting rid of waste—are carried out.

9–12 The Cell

- Within the cells are specialized parts for the transport of materials, energy capture and release, protein building, waste disposal, passing information, and even movement.

Related Research

- Studies on children's understanding of life science have revealed that students harbor many alternative conceptions relating to basic biological concepts even after instruction (Driver et al. 1994).
- In a study where high school biology students were given a question that asked them to identify which things were an excretory process, most students did not view exhalation as an excretory process. The researcher concluded that one possible reason students were not able to relate the removal of carbon dioxide during exhalation to an excretory role is that exhalation is learned in the context of breathing and is not linked with excretion (Din-Yan 1998).
- The erroneous view that undigested food waste is removed by the excretory system was most prevalent among average high school biology students. This view suggests that the idea of metabolic waste is a difficult and abstract concept for the average student and is not well understood even after formal instruction (Din-Yan 1998).
- In Din-Yan's study (1998), some students wrongly selected that the release of saliva was an excretory process. This may indicate that some students tend to consider that secretions made by the body usually contains some unwanted materials, which does not hold true for saliva. This reflects some confusion about the nature and roles of secretion and excretion in biological processes.
- In a follow-up interview after students were asked to choose things that are removed by the excretory system, many students could define *excretion* and *egestion* correctly, but further questioning revealed that they simply memorized the definitions and the examples they gave for excretory substances (such as feces and saliva) were incompatible with their definitions (Din-Yan 1998).

Suggestions for Instruction and Assessment

- Removal of waste products is a function of all living things. Before developing the formal concepts of *excretion* and *egestion*, you should first assess students' understanding of "waste products." Have students generate a list of "wastes" removed by human body processes. Then challenge students

★ Indicates a strong match between the ideas elicited by the probe and a national standard's learning goal.

to think about which wastes are produced within cells as a result of metabolic processes and which ones are not produced within cells.

- When teaching about the respiratory system, explicitly point out the difference between *exhalation* as part of the ventilation process that forces carbon dioxide out of the lungs and *excretion* as the process that removes carbon dioxide as a metabolic waste resulting from cellular respiration.
- Compare everyday meanings of *excretion* and *elimination* with their biological meanings.
- An effective way to help students consider the connections among different metabolic processes and excretion is to develop concept maps (Novak and Gowin 1984). Develop a concept map as an instructional tool or have biology students develop concept maps that link excretion to metabolic processes such as cellular respiration, breakdown of proteins, removal of excess mineral salts in the blood, and removal of bile components.
- Help students see the connection between the excretory system and the circulatory system, further reinforcing the big idea of how one system is connected to another system. CO_2 and urinary wastes should be linked by the idea that both are excreted from cells into the *blood*. Kidneys filter the urinary waste out, while CO_2 diffuses out into the atmosphere when carried to the lungs. Both are picked up and dropped off by the circulatory system.
- Be aware that many traditional textbooks incorrectly describe excretion as the removal of all waste products, both metabolic and nonmetabolic. Often, the way human body systems are presented in textbooks fails to make a distinction between elimination of waste materials as a general function and excretion of metabolic wastes specifically. Some textbooks are not explicit enough when they show the connections between the digestive, respiratory, and excretory systems as systems that get rid of wastes and the different types of wastes removed in these systems, as well as how they are removed.
- In animals, the processes of solid, liquid, and gaseous waste removal are intimately connected. Help students recognize that the routes for elimination of undigested materials and excreted wastes can differ between animal species. For example, in birds and reptiles nitrogenous waste is eliminated as part of fecal matter, not as urine.

Related NSTA Science Store Publications, NSTA Journal Articles, NSTA SciGuides, NSTA SciPacks, and NSTA Science Objects

American Association for the Advancement of Science (AAAS). 2001. *Atlas of science literacy.* Vol. 1. (See "Cell and Organs" map, pp. 74–75.) Washington, DC: AAAS.

Science Object: *What Happens to the Food I Eat?*

Related Curriculum Topic Study Guides (in Keeley 2005)
"Cells"
"Human Body Systems"

References

American Association for the Advancement of Science (AAAS). 2009. Benchmarks for science literacy online. *www.project2061.org/publications/bsl/online*

Din-Yan, Y. 1998. Alternative conceptions on excretion and implications for teaching. *Education Journal of the Chinese University of Hong Kong* 26 (1): 101–116.

Driver, R., A. Squires, P. Rushworth, and V. Wood-Robinson. 1994. *Making sense of secondary science: Research into children's ideas.* London: RoutledgeFalmer.

Keeley, P. 2005. *Science curriculum topic study: Bridging the gap between standards and practice.* Thousand Oaks, CA: Corwin Press and Arlington, VA: NSTA Press.

National Research Council (NRC). 1996. *National science education standards.* Washington, DC: National Academies Press.

Novak, J., and D. Gowin. 1984. *Learning how to learn.* Cambridge, UK: Cambridge University Press.

Antibiotics

William has a cold. A cold is caused by a virus. William wants to get better soon so he can play in the basketball tournament. His mother calls the doctor and asks for antibiotics. Do you think antibiotics could help William get over his cold quicker?

Circle your response:

Yes No Maybe

Explain your thinking. Describe your ideas about taking antibiotics for a cold.

Antibiotics

Teacher Notes

Purpose

The purpose of this assessment probe is to elicit students' ideas about infectious diseases. The probe is designed to reveal whether students recognize that antibiotics only work on bacterial infections, not viral infections.

Related Concepts

Infectious disease, antibiotics, virus

Explanation

The best answer is no. Antibiotics are not used to treat the common cold, or any kind of infectious disease caused by a virus (e.g., the flu). Antibiotics are used to treat infectious diseases caused by bacteria. There are more than 100 different antibiotics. The first antibiotic, penicillin, was discovered accidentally by Alexander Fleming in 1928 from a mold-contaminated bacterial culture. Common antibiotics include penicillin, amoxicillin, erythromycin, tetracycline, and sulfa drugs. Antibiotics work by blocking the growth and reproduction of bacteria, not viruses such as the common cold. They are like a selective bacterial poison used to kill the bacteria but not human cells. Different antibiotics are used for different types of bacterial infections, depending on the bacterial agent causing the infection.

All antibiotics work by killing or impairing the bacteria, and they do this in a variety of ways. For example, an antibiotic might inhibit a bacterium's ability to turn glucose into energy, and without energy the bacteria cannot live. Some prevent the manufacture of a necessary protein, resulting in death of the bacteria. Some antibiotics affect the bacterium's cell wall, preventing cell division and compromising the protection of the bacterium, causing it to die. Others weaken the bacteria so the immune system can take over.

The important thing to recognize is that because viruses are not living and therefore do not need energy, produce proteins, reproduce on their own, or carry on any of the other life processes that antibiotics interrupt, antibiotics

are useless against viruses. However, antibiotics are one of the most prescribed medications and are often prescribed for viruses, even though they do not affect viruses. This misuse of antibiotics is one of the contributing factors to bacterial resistance. Every time a person takes antibiotics, sensitive bacteria are killed, but resistant bacteria may be left to grow, multiply, and pass on their genes. Repeated and improper uses of antibiotics are primary causes of the increase in drug-resistant bacteria.

Another negative consequence of using antibiotics for viruses is that antibiotics can't discriminate between bacteria that are good for us and bacteria that cause disease. We coexist with a wide variety of helpful bacteria. For example, our intestines are lined with bacteria that break down foods that we can't digest. Whenever you take antibiotics, you kill off some of these good bacteria. Using antibiotics indiscriminately can destroy much of the bacteria normally found in your body, providing an opportunity for harmful strains to establish themselves in their place.

Curricular and Instructional Considerations

Elementary Students

In the elementary grades, students learn about "germs," and generally do not use the words *bacteria* and *virus*, or know the difference between them. They often attribute all diseases to germs without distinguishing between infectious and noninfectious diseases. Elementary students begin to understand the idea of contagious diseases and the importance of hand washing and proper sanitation to reduce the spread of disease. Although they may not know what an antibiotic is, they are familiar with antibiotics from everyday experiences of having had ear or other types of infections for which they or someone in their family may have had to take antibiotics.

Middle School Students

By middle school, students recognize the various factors that contribute to disease—microorganisms, genetic makeup, personal health habits such as nutrition, injury to or malfunction of organs and organ systems, and the environment. They can distinguish between infectious and noninfectious diseases and recognize how their immune systems respond to infections. They have the vocabulary for many aspects of health but often do not understand the science related to the vocabulary (e.g., antibiotics). Developing a scientific understanding of health is the focus of the "Science in Personal and Social Perspectives" standards in the National Science Education Standards (NRC 1996), which distinguish learning goals in science from the topics students learn in health class. Middle school students are introduced to and become interested in the historical discoveries that led to the formation of the germ theory of disease, including Fleming's discovery of penicillin.

High School Students

By high school, most students have an understanding of the structure and function of major human body systems such as the digestive, respiratory, and circulatory systems. However, they may not have as clear an understanding of other systems, such as the nervous, endocrine, and immune systems. Therefore, students may have difficulty with specific mechanisms and processes related to health issues (NRC 1996). High school students distinguish between prokaryotes and viruses and recognize that bacteria have cell membranes and viruses do not, one prerequisite to understanding how an antibiotic works on bacteria but not viruses. They should know how a variety of infectious diseases are caused by bacteria, viruses, or parasites. However, through their everyday experiences with colds, flu, and other infectious diseases, students may believe that antibiotics

are prescribed for all infectious illnesses. At this level students expand their knowledge of medical technology and the science that makes the discovery and synthesis of powerful antimicrobial drugs possible.

Administering the Probe

This probe is best used with middle and high school students. Make sure students are familiar with the word *antibiotic* before giving them this probe. You might start off by asking if students have ever been given an antibiotic when they were ill and what type of antibiotic they received. Elicit responses such as penicillin or other types of antibiotics. Make sure students know that a cold is caused by a virus. (*Note:* This is stated in the probe.)

Related Ideas in *National Science Education Standards* (NRC 1996)

K–4 Personal Health

- Individuals have some responsibility for their own health. Students should engage in personal care—dental hygiene, cleanliness, and exercise—that will maintain and improve health. Understandings include how communicable diseases, such as colds, are transmitted and some of the body's defense mechanisms that prevent or overcome illness.
- Students should understand that some substances, such as prescription drugs, can be beneficial, but that any substance can be harmful if used inappropriately.

5–8 Structure and Function in Living Systems

- Disease is a breakdown in structures or functions of an organism. Some diseases are the result of intrinsic failures of the system. Others are the result of damage by infection by other organisms.

9–12 Personal and Community Health

- The severity of disease symptoms is dependent on many factors, such as human resistance and the virulence of the disease-producing organism. Many diseases can be prevented, controlled, or cured.

9–12 The Cell

- Cells have particular structures that underlie their functions.

Related Ideas in *Benchmarks for Science Literacy* (AAAS 2009)

K–2 Physical Health

- Some diseases are caused by germs, some are not. Diseases caused by germs may be spread by people who have them. Washing one's hands with soap and water reduces the number of germs that can get into the body or that can be passed on to other people.

K–2 Health Technology

- Medicines may help those who do become sick to recover.

3–5 Physical Health

- Some germs may keep the body from working properly. For defense against germs, the human body has tears, saliva, and skin to prevent many germs from getting into the body and special cells to fight germs that do get into the body.

6–8 Physical Health

- Viruses, bacteria, fungi, and parasites may infect the human body and interfere with normal body functions. A person can catch a cold many times because there are many varieties of cold viruses that cause similar symptoms.

- Specific kinds of germs cause specific diseases.

6–8 Health Technology

- Many diseases are caused by bacteria or viruses.
- ★ If the body's immune system cannot suppress a bacterial infection, an antibacterial drug may be effective—at least against the types of bacteria it was designed to combat. Less is known about the treatment of viral infections, especially the common cold. However, more recently, useful antiviral drugs have been developed for several major kinds of viral infections, including drugs to fight HIV, the virus that causes AIDS.

9–12 Health Technology

- Knowledge of molecular structure and interactions aids in synthesizing new drugs and predicting their effects.
- The incorrect use of any given antibacterial drug can lead, by means of natural selection, to the spread of bacteria that are not affected by it.

Related Research

- Many elementary students believe that all diseases are caused by the same kind of germ (AAAS 2009).
- Studies have found that antibiotics are a mysterious concept to the general public, including students (Lucas 1987 and Prout 1985). In almost all sample groups questioned, most respondents did not know that antibiotics act only on bacteria and not on viruses. Many appeared to confuse antibiotics with antibodies, and it was further found that this was not just confusion about similar-sounding words. Even people speaking a language in which the two words were not similar in sound were also confused about the difference between the two (Driver et al. 1994).
- In a survey commissioned by the Royal Pharmaceutical Society of Great Britain (RPSGB), 55% of the 250 respondents wrongly thought antibiotics could cure all coughs and 47% that they could cure all cases of flu and colds (BBC 1999).
- In a Gallup poll of more than 1,000 people (all nonphysicians), 60% responded that antibiotics were effective against cold and flu (Udall 1997).

Suggestions for Instruction and Assessment

- Combine this probe with "Catching a Cold" from Volume 4 (Keeley and Tugel 2009). This probe addresses another myth associated with viral diseases.
- Help students understand that the word *antibiotic* means against (anti) living (biotic). Antibiotics work against living organisms that invade our bodies, specifically bacteria. Because viruses are not considered living, an antibiotic would not affect them. Also, antibiotics are specific to different types of bacteria. They do not work on other living infectious agents such as fungi, protists, and animal parasites.
- Be aware that some doctors still prescribe antibiotics, at a patient's insistence, for illnesses that are not bacterial. When these viral illnesses clear up, it is not because of the antibiotics, but because the viral infection has run its course as a result of the immune system taking over. It is important to make students aware of this overuse of antibiotics and how it contributes to bacterial resistance. (*Note:* Some physicians prescribe antibiotics during a viral infection to treat secondary bacterial infections that may arise when the immune system is weakened.)

★ Indicates a strong match between the ideas elicited by the probe and a national standard's learning goal.

- For older students, connect the way antibiotics work with life processes carried out by cells. By disrupting bacteria's life processes, the infection is treated.
- Older students may be interested in learning how various microbiological techniques identify which type of antibiotic to use. For example, the Gram stain is based on the properties of the bacterial cell wall and is used to identify certain groups of bacteria. A gram-positive stain (purple) or a gram-negative stain (pink) provides important diagnostic information as to the type of organism and antibiotic that could treat the infection.
- Connect this probe to the historical episode that led to the discovery of antibiotics (Alexander Fleming's mold culture) as well as the historical discoveries that led to the germ theory of disease, leading up to our present-day health technologies and medical engineering. The story of modern medicine illustrates some of the greatest advances in the dual roles of science and technology.

Related NSTA Science Store Publications, NSTA Journal Articles, NSTA SciGuides, NSTA SciPacks, and NSTA Science Objects

American Association for the Advancement of Science (AAAS). 2001. *Atlas of science literacy.* Vol. 1. (See "Disease" map, pp. 86–87.) Washington, DC: AAAS.

Related Curriculum Topic Study Guides (in Keeley 2005)
"Infectious Disease"
"Health and Disease"
"Medical Science and Technology"

References

American Association for the Advancement of Science (AAAS). 2009. Benchmarks for science literacy online. *www.project2061.org/publications/bsl/online*

British Broadcasting Corporation (BBC). 1999. Antibiotic confusion widespread. *http://news.bbc.co.uk/2/hi/health/471683.stm*

Driver, R., A. Squires, P. Rushworth, and V. Wood-Robinson. 1994. *Making sense of secondary science: Research into children's ideas.* London: RoutledgeFalmer.

Keeley, P. 2005. *Science curriculum topic study: Bridging the gap between standards and practice.* Thousand Oaks, CA: Corwin Press and Arlington, VA: NSTA Press.

Keeley, P., and J. Tugel. 2009. *Uncovering student ideas in science, Vol. 4: 25 new formative assessment probes.* Arlington, VA: NSTA Press

Lucas, A. 1987. Public knowledge of biology. *Journal of Biology Education* 21 (1): 41–45.

National Research Council (NRC). 1996. *National science education standards.* Washington, DC: National Academies Press.

Prout, A. 1985. Science, health, and everyday knowledge. *European Journal of Science Education* 7 (4): 399–406.

Udall, K. 1997. *Nature's antibiotics.* Provo, UT: Woodland Health Publisher.

Index

Index

Index

Index

Index

Index